英汉双语云南少数民族经典文化概览丛书

丛书主编◎李强

Different Clothing Culture of Yunnan Ethnic Groups

风格迥异的服饰文化

曹霞　编译

云南出版集团

云南人民出版社

图书在版编目（CIP）数据

风格迥异的服饰文化：汉文、英文／曹霞编译. ——
昆明：云南人民出版社，2022.8
　　（英汉双语云南少数民族经典文化概览丛书／李强
主编）
　　ISBN 978-7-222-21121-6

　　Ⅰ. ①风… Ⅱ. ①曹… Ⅲ. ①少数民族—民族服饰—
服饰文化—介绍—云南—汉、英 Ⅳ. ①TS941.742.8

中国版本图书馆CIP数据核字(2022)第139650号

责任编辑：陈　晨
装帧设计：张　艳
责任校对：任建红
责任印制：窦雪松

英文校对：杨　燕

英汉双语云南少数民族经典文化概览丛书

风格迥异的服饰文化
FENGGE JIONGYI DE FUSHI WENHUA

丛书主编　李　强
曹　霞　编译

出　版　云南出版集团　云南人民出版社
发　行　云南人民出版社
社　址　昆明市环城西路609号
邮　编　650034
网　址　www.ynpph.com.cn
E-mail　ynrms@sina.com
开　本　720mm×1010mm　1/16
印　张　9
字　数　210千
版　次　2022年8月第1版第1次印刷
印　刷　昆明德厚印刷包装有限公司
书　号　ISB N978-7-222-21121-6
定　价　52.00元

云南人民出版社微信公众号

如需购买图书、反馈意见，请与我社联系
总编室：0871-64109126　发行部：0871-64108507　审校部：0871-64164626　印制部：0871-64191534

序　言

　　服饰是人类社会文明进步的标志之一，服饰是紧跟时代而变化的。本套系列丛书以中华民族文化自信为经，以中国少数民族经典文化国际传播为纬，全面系统地据实介绍云南少数民族社会生活特征和独特的文化传承模式。此套丛书也是我院在"十四五"开局之年，依托多语文化比较研究中心而实施的学科建设发展项目的重要组成部分。此书《风格迥异的服饰文化》的编译者曹霞副教授是我院优秀的青年骨干教师。

　　《风格迥异的服饰文化》一书作为系列丛书的成果之一，在语言格式上采用英汉双语对照模式，以满足中国优秀文化经典要素国际传播的实际需求，更好地服务于国家"一带一路"建设。服饰是人类生活的要素，也是一种文化载体。在中华上下五千年的文明史中，服饰承载着厚重的传统文化内涵，它不仅可以驱寒蔽体，而且满足了人们的审美情趣，反映了民俗风情和社会制度，记录了历史和社会生活，体现了人们的思想观念和民族精神。一套民族服饰就是一个民族的缩影。在一定意义上，此书的问世，对涉外工作人员、大专院校学生和海内外友好人士深入了解云南民族多元传统文化内涵具有重要价值。让我们一同走进异彩纷呈的云南民族服饰世界，感叹霓裳羽衣的绚丽多姿，品鉴神奇的云南民族服饰文化。

　　为此而言，未能述全，当作序。

云南民族大学外国语学院多语文化比较研究中心

李骞

2022年8月于昆明呈贡大学城

目　录

Contents

第一章　悠久的历史　千姿百态的服饰

　　云南传统服饰文化以其众多的服型、多姿的款式、缤纷的色彩及其独具特色的文化内涵，已经构成了一个自成体系的文化类型。在漫长的发展过程中，云南少数民族服饰融入了一些特殊的文化内涵。在看似简单的衣装中，融入了一些鲜活而生动的文化内容。因此，服饰不再是简单的衣装，而是一种包含着丰富文化内容的载体，故我们称之为"服饰文化"。

　　云南民族服饰文化可用"多""丰""彩""巧"来形容。"多"，一是指民族多，二是指服型多。"丰"是文化内涵积存丰厚。"彩"即色彩丰富、斑斓多姿。"巧"，即构图严谨、疏密有致、巧夺天工。

　　云南是我国少数民族数量最多的一个省份，世居少数民族达25个。单是云南特有的少数民族就达15个。这些民族还有众多的分支，即便是同一个民族，其分布地也较为广泛。各地区、州、县（市）几乎都有少数民族分布。省内各地几乎没有独立分布的单一民族，一般与一个或数个民族相邻或交错而居，在同一地区各民族分布又多呈立体分布。以西双版纳州勐海县为例，傣族多居坝区和河谷地带。哈尼族、布朗族、拉祜族等族则居住在半山区和山区。云南各族所在的经纬度位置及气候带也呈多样性。傣族、阿昌族、德昂族等族多居热带，而藏族、纳西族、普米族等民族则居高寒山区。

　　另外，各民族的社会、历史、经济及文化发展也不平衡。到20世纪50年代初，云南少数民族或处于封建半封建社会、奴隶社会，有的民族甚至还处于原始社会后期。这些差异和不同，造就了云南少数民族服饰文化的多样性，呈现出纷繁复杂和多姿多彩的格局。在服饰文化中，服型和款式是两个重要的文化元素。云南民族服饰以其多样的服型和众多的款式闻名全国。从与服饰起源有关的单件衣装到成套衣装，一直到华丽庄重的土司服装及其饰品，构成了文化内涵丰富、博大精深的云南服饰世界。

　　在工艺方面的成就也是值得世人称道和赞叹的。多种工艺集于一身，是云南服饰文化中惯用的技法。云南的少数民族喜欢采用刺绣、扎染、蜡染等多种工艺，对自己的衣装进行巧妙的装饰。其中，刺绣工艺是最为突出的云南民族服饰的主流工艺。

在服饰美学方面，也可找寻到它的独到之处。在或朴素或奇美的服饰图案中，大多被赋予了丰富的文化内涵，这是云南少数民族服饰文化中最为生动、最具魅力，也是最具生命力的文化要素。

所谓"一身穿戴，多种信息"正是对云南少数民族服饰所具有的丰富文化内涵最为生动而贴切的描述。

云南悠久的历史，众多的民族，复杂的地形地貌和多变的立体气候，决定了云南少数民族的服装和装饰必然绚丽多彩、蔚为大观。云南有26个人数较多的民族。各民族又以文化与经济等方面的差异而分出许多支系。民族不同，服装不同，支系不同，服装也有差别，甚至同支系而居地不同，服饰也会出现差异。云南的民族服饰，因此也就纷繁复杂，数量极多。

远在170万年前，便有旧石器时代早期的直立人——元谋人在云南居住，旧石器时代中期的早期智人昭通人，旧石器时代晚期的晚期智人丽江人、西畴人、昆明人、蒙自人等在云南居住。旧石器时代人类的生产方式是实行掠取经济：狩猎、采集、捕捞。人类的居所则多选择天然洞穴。人类为了御寒，使用的"服装"是树叶、树皮、鸟羽、兽皮、兽毛等。

进入新石器时代，人类进入生产经济时代。人类学会了种植谷物、驯养畜禽、制陶和纺织。由于纺织技术的发明，人类开始织麻制衣以御寒蔽体。云南新石器时代遗址中发现的陶纺轮，便是人类纺织制衣的工具。新石器时代的云南原始居民十分注意装饰自己的身体。人们用骨、角石等材料制成装饰品装饰身体各部位。在沧源崖画上，就有许多人物的头、颈、耳部用兽角、羽毛、树枝加以装饰。此外，云南新石器时代的居民还喜爱一种永久性装饰——文身。沧源画上就有文身人像。

进入青铜时代，云南古代民族的服饰的多样化和个性化体现在青铜器人物图像上。通过不同的服饰、发型，可以识别出他们属于古代何种民族。在《史记·西南夷列传》中，司马迁记载了滇人的穿着。滇人无论男女，都穿一种无领的对襟上衣，长至膝，下体无裤无裙。

到了唐宋时期，由于统治云南的是南诏国和大理国这两个地方政权，君臣上下，等级森严，服饰也非常讲究。唐人樊绰的《蛮书》记载了很多这方面的资料。

据记载，当时南诏国的国王和宰相所穿的衣服都是有彩色花纹的丝织品。贵妇人穿罗锦缎制的裙子，发髻上装饰珍珠、金贝壳、琥珀。

元、明、清时期，云南各少数民族的服饰已呈现出百花齐放的特色，尤其是清代云南各民族服饰奠定了云南近现代民族服饰的基础。

深厚的历史文化积淀与多民族文化的格局，成为今天丰富灿烂的云南民族服饰文化的基础。

同时，云南又是东南亚、东亚与南亚之间的交通要冲。很早就有一条经四川过云南到印度的贸易通道，贸易往来十分频繁。《蛮书》与《马可·波罗游记》之类，都对此有详细的记载。云南人本身，在国外的贸易与内地的商业贸易中，也采取主动、积极的态度，一向活跃。

云南的许多民族，本身也经历了长期的迁徙历程，从四面八方汇聚在云南这片土地上。不同的生活经历与迁徙历程，把许多不同的文化带入云南，极大地丰富了云南的民族文化。当初作为外来文化的汉文化，进入云南后，不仅扎根在汉族之中，也影响到少数民族地区，形成自己的独创风格。

云南复杂的地理环境、多样的气候与丰饶的物产，激发了云南各族人民丰富的创造力。云南山河纵横、气候多样，从低海拔的热带雨林到6000米以上海拔的雪山都有。人的生活因地理与气候等因素的不同而发生适应性的变化，这在服饰上表现得尤其明显，迪庆藏族适应高海拔寒冷地理条件的皮毛服装与西双版纳傣族的轻便服装，就是最好的例子。一般来说，冷凉地带的民族多着长宽厚重的服装，而温热地带的居民则可以较少考虑保暖而偏向轻薄，中间地带也依条件调适，形成很多独特的服饰风格。

云南丰富的物产，使各族人民展示创意，做出格调迥异的服饰。棕榈、山草、树皮、树皮与树叶等用作衣饰材料，丰富了材料来源。巨大的鸟嘴壳、美丽的孔雀、锦鸡尾羽与虎爪兽牙等用作饰品，则增添了奇特的装饰效果与风俗文化。云南极多的饰品，还仰赖云南本土丰富的金、银、铜、锡与玉石资源。

正因为汇聚了众多的优势，云南的民族服饰不仅种类繁多，文化沉积也十分深厚。各族儿女往往把自己对美好事物的感受，走过的路与经历过的事，遵循的社会道德，都用服饰文化表现出来。一片片的绣件，一件件的服装，处处可见各族人民质朴的审美思想与深沉的生活哲学。

第二章　民族服饰概览

云南民族众多，服饰千姿百态。由于特殊的地理环境和历史原因，居住在这里的25个世居少数民族，各自都有自己独特的服饰装束。有的民族因支系不同，服饰也不同，即使是同一支系，也往往因居住地域的差异而服饰各有千秋。更重要的是，各民族的服饰，在历史发展的长河中，各自都融进了特殊的文化内涵。各民族的服饰，就是一幅幅独具风情趣味的画卷，也是一座座民族民间工艺美术的宝库。云南少数民族服饰是民族生活和生产方式，以及风俗习惯、审美情趣的真实写照，也是民族文化的直观体现。云南的少数民族服饰色彩斑斓、风情万种，下面选取一些特色鲜明的少数民族服饰加以介绍。

佤族服饰

佤族主要聚居于临沧市沧源佤族自治县和普洱市西盟佤族自治县。女裙以五色线织成主色为红色，间杂他色；由骨牌和五色料珠串连而成的腰带和银头箍是其服饰的根本标志。

佤族服饰特点突出，头饰、身饰与腰箍等，都具有典型的民族风格。佤族服饰给人的总体感觉是结构与花式较为简单，是用简单的线条与构图创造自己鲜明的特征。佤族服饰的颜色以黑色为主调，只有耿马傣族自治县有少数佤族衣带黄土色，沧源自治县与西盟自治县是佤族服饰的主要流行区。部分佤族男子戴耳环，过去多数佩刀。双江拉祜族佤族布朗族傣族自治县的男子大裤，裆可垂至踝，极具特色。

佤族服饰

布朗族服饰

布朗族主要聚居于西双版纳傣族自治州勐海县布朗山乡和打洛镇，是一个跨境民族。其服饰深受傣族的影响，但也自成一格，其中织纹筒裙最具代表性，饰品以银耳柱独具民族特色。

布朗族妇女服饰差别较大的是居住在西双版纳州和澜沧江一带的布朗族女青年，上着圆领窄袖开襟短衫，下着两层筒裙。姑娘留长发，妇女则梳发髻，上插簪针，针顶镶嵌三颗菱形透明玻璃珠，下系一条银链，垂吊银片、银铃等饰物。施甸地区的妇女服饰，别具一格。上身穿高领、长袖，大面斜襟衣，袖口镶红绿花布横条，高领上绣精美的花纹图案，颈戴银泡项链，外罩一件花布对襟短褂，襟边镶钉15～20对银币小纽扣，胸前系滚白边的青布围腰长至膝部，下身着青布长裤，扎青布绑腿，挽髻用3米多长的两块青布包头，包头折叠成三角状，用一条彩色玻璃珠穗扎紧，插一朵白绒球，足穿绣花鞋。

　　布朗族人数较少，多与傣族杂居，受其影响较大。因居住分散，各地也有差异。

　　（1）勐海式。主要分布在勐海县、景洪市、澜沧拉祜族自治县等地。主要有开襟长袖小衣、长筒裙、腰裙、无领凹口对襟短褂、包头与右衽长袖衣、绑腿等。三尾螺头簪是最具代表的饰品。

　　（2）双江式。主要在双江拉祜族佤族布朗族傣族自治县。包头，斜衽右襟衣等有特色。

　　（3）云县式。旧式服装有特色，今已变异，同社会一般服饰。

　　（4）思茅式。近水傣服饰。

　　（5）耿马式。同当地傣族服饰。

德昂族服饰

　　德昂族主要分布在德宏傣族景颇族自治州、保山市、临沧市与普洱市的澜沧拉祜族自治县。德昂族妇女黑布包头，穿藏青或黑色对襟上衣，襟边镶有直条红布，并钉上大方块银扣4～5副。下穿筒裙，打绑腿，缠腰箍。德昂族妇女筒裙较长，上遮乳部，下长及踝骨，裙身织有色彩鲜艳的横条纹。因其条纹和底色不同而有"花德昂""红德昂"和"黑德昂"之称。"花德昂"的筒裙上是红、黑相间的宽粗横条纹，也有的镶上四条白带，白带中再插入17厘米宽的红布条；"红德昂"的筒裙上织有宽约17～20厘米的大红横条；"黑德昂"的筒裙用黑线织成，其上间织红、绿、白色细条纹，裙的长度从腰部至踝骨，比其他两个支系的裙短。

　　德昂族妇女腰部套着五六个甚至二三十个腰箍，腰箍用藤或篾制成，多漆成红、黑色，有的还刻有各种花纹，多为男方追求女方时，由男方圈套在女方腰上，作为两情相悦的标记，腰箍是德昂族妇女服饰中的特色装饰，其文化意蕴也非常古老。

傣族服饰

傣族服饰，可分为"傣泐"（水傣支系）、"傣那（旱傣支系）和"傣雅"（花腰支系）三个类型。

"傣泐"支系，以西双版纳为主，妇女服饰以清雅纯净为特点，少女挽髻于顶，稍偏右侧，髻上插鲜花或彩色发梳。上身内穿无领无袖紧身衣，有似胸罩功能。外穿右襟圆领短褂，长仅过脐，袖窄长，紧束肘臂。上衣色彩多为纯红、粉红、水绿、赤黄等。下着筒裙，裙用彩色布制成，裙长及脚面。束银腰带，以带宽和纹细为美。腰带束于短衫之下，赤足或穿凉鞋。出门肩拎"筒帕"（挎包），喜戴耳环、手镯、项链等。

"傣那"型服饰，以德宏为主，包括保山、耿马、临沧等地区。少女头部用红线绳结发辫盘绕于顶，再插入金银饰品，常戴篾制小帽。上身穿白色或浅蓝色右襟短衣，胸前佩金银质的龙牌饰或花朵等物，下身不着筒裙而穿黑色长裤。束青色绣花小围腰，围腰佩有长长的飘带打结垂于右侧，长至脚踝。有的还披色彩艳丽的披巾。

婚后，妇女头戴黑布缠成的高筒帽，帽边用绿色头绳缠绕为饰。上着对襟短衫，下着黑色筒裙，膝下至踝处用青布缠裹，便于劳作。

"傣雅"型服饰，主要分布在新平、元江两地。妇女上穿花边紧胸短背心，仅及胸，胸襟用蓝色土布或粉红、草绿色布制作，下端钉一排细银泡。外穿黑色短衫，袖长而窄，衣领上饰银泡，衣身镶花边。下身着黑色筒裙，质地较厚，长至膝与踝之间，下摆饰6厘米宽的花边。筒裙外束围腰共3条，从里到外，一层比一层短，显出每层花边，最里一层长至膝部。系挑花腰带，腰带长1米左右，上绣植物、几何纹样。另有银质腰带两条：一条为带形，镶的银泡较大，两端挂着银穗，上接内衣衣襟处，至围腰正面左下端；另一条为三角形，上饰银泡较小，下沿垂丝线，长至裙边，束此腰带能将整个臀部覆盖。外出均在后腰间挂上笆箩，笆箩为竹编，做工精细，上饰毛线花。

彝族服饰

彝族是服饰文化积存最为丰厚的民族。仅就其服装款式，就能分为：大小凉山、滇西、滇中、滇南、滇东北、滇西北六大类型，每一类型中又都不下几十种款式，每种款式都有它的特色。

其中，大小凉山男子，头饰"天菩萨"，身披"察尔瓦"的装束，文化渊源非常古老，而滇中、滇南妇女的服饰，挑花绣朵，极其精细，反映出刺绣工艺的高超水平。

滇中，以楚雄地区为例，妇女上穿短衣、下着长裤，系围腰，围腰有兜肚形和方形两种，均绣精美图案。妇女头饰繁多，大体可分为包帕缠头、戴绣花帽等。上衣多采用桃红、翠绿、碧蓝布做成，极少有黯淡的色彩出现。多在领口、托肩、袖口、裤脚边刺绣花卉图案。有的还在环肩部位镶嵌一圈花边或缀以彩色丝线流苏，把花团锦簇的浓艳特色体现得更加淋漓尽致。特别是大姚县昙华山一带绣制的挎包，图案古朴典雅，工艺精巧细密，与男女服饰相配，更是多彩多姿。

滇南，以红河地区为代表。其中石屏县"花腰彝"妇女服饰是挑花刺绣中最为精细奇美的典范。其上穿黑色或青蓝色紧身长袖大襟长衫，衫长及膝，肩、下摆与袖口有刺绣图案和色布装饰，外罩一件形制独特的对襟坎肩，坎肩几乎是用精细的挑花做成，襟边也有约10厘米的挑花图案和银泡镶嵌，银币为扣。头戴顶巾帽，帽用长80厘米、宽约40厘米的绣花巾盘结而成。下着长裤，往往将长衫后襟撩起，或翻叠于腰后，底襟边饰则显露出来。腰系色彩华丽、工艺精致的绣花腰带。石屏彝族妇女服饰，以坎肩和腰带最为别致。

哈尼族服饰

哈尼族服饰类型也很多，其中，以"爱尼""白宏""奕车三个支系的服饰最具特色。爱尼，集中居住在西双版纳和普洱市。妇女头戴竹帽，称作"吴确"，有圆形和三角形两种，帽内衬有黑布，帽的前部用奶色草秆皮裹就的圈上钉数十颗银泡，再围彩色珠串和色线。少女头戴有银泡的小帽，上插红色羽毛，有的还饰有牛角做成的骨针和昆虫等。

妇女内穿黑色或青色胸衣，长约30厘米，镶布条有菱形、圆形银牌和数行小银泡，胸衣仅遮住乳部，用一条背带斜挎于肩。外套圆领右襟上衣，背部有三角纹、方格纹、波形纹的丝线挑花图案，袖口也有红、白、蓝色布块镶缀。下穿长不过膝的短裙，系于髋骨上，只遮住臀的下半部，已婚者系得更低。腰系串着海贝的彩色腰带，带的两端下垂有彩色珠链，颇为俏丽。挎包是女子的主要佩带物，用白色和黑色的土布做底，四周镶上花布条，中间绣上图案，钉上银泡、银币，再上绒球，使爱尼女子服饰更添秀色。

白宏支系的妇女服装以短、紧、小为特点。上衣长不过脐，用银币作组扣，但从不扣上，以能露小腹和脐为美。在上衣的前胸部位钉六排银泡，正中一枚八角形大银片，犹如一朵盛开的白莲，下穿双折短裤，长不过膝，裹绣花绑腿。少女戴镶银泡的小圆帽，成婚生育后戴蓝布包头。整套服装不失现代女性健美的韵味。

　　奕车支系的妇女服饰与其他支系的相比，特色更浓。妇女上装分3层，大多用自织自染的土布缝制，最里面是背心，中间是衬衣，外面为外衣。外衣共有9层之多，统称为龟服。无领、开胸、紧身、半截，穿上不但手臂外露，而且右边胸也半露着，只将左边胸遮掩严实。前胸佩带银链，与素雅简朴的藏青色衣服配套，十分典雅。奕车姑娘腰间紧束一条宽约10厘米的绣花腰带，前面缀上银币、流苏，不仅把姑娘身体的曲线勾勒得十分柔美，而且显示出少女的干练和英气。腰带外又系一串10多个螺蛳形状的银腰链，走起路来叮当作响，下装仅穿一条青色短裤，并赤足。从脚掌到整个大腿根部全都外露，数九寒天也不例外。短裤在剪裁上十分考究，一般由姑娘根据自己身材亲自缝制，以能紧束臀部为佳。短裤下端底边要呈人字形，对折出七道折子来，看上去像穿了七条重叠的短裤。最为显眼的是每位奕车女子都喜欢随身带一把白色土布伞，与黑色服饰相衬，十分显眼。

　　奕车女子的头饰，为一种三角形的白色尖顶头布，形状似雨衣上的雨帽，但后面多了一对好看的燕尾，并在头巾边沿绣上精美的彩色花纹。这种头饰，与古老的传说有关，有着特殊的服饰文化内涵。

白族服饰

　　女装，小领或无领右衽内衣，长及胯或膝，年轻者多着白、浅蓝色，中年以上偏于蓝、黑色。外罩红或浅蓝色坎肩，右衽结处挂"三须"或"五须"银饰，腰系绣花方围腰，臀垂飘带，下着浅色宽脚裤，足穿绣花鞋；用挑绣或印染方帕及彩色毛巾包，未婚女编独盘于顶，已婚妇女改为挽髻。剑川一带白族未婚姑娘则戴小帽或布满玉兔银泡的"鼓钉帽"或"鱼尾帽"。

　　男子多穿白或浅蓝色上衣，外罩黑布或白羊皮领，下穿扭裆宽脚，绣花缀缨布凉鞋或剪刀口扣绊布底鞋。白色或浅蓝色布包头，垂下尺许。洱海边渔民喜戴瓜皮小帽，穿组扣多且短得露脐的上衣。层数越多越好看，腰系红绿带，挂荷包，戴麂皮或布质花肚兜。衣服穿的层数多，称"千层荷叶"。穿三层内长外短的称"三叠水"，视此为俊美、富足。

纳西族服饰

　　纳西族妇女穿右大襟肥袖衣，外套一件前襟短后襟长的无领夹褂，腰宽袖大，袖长及臂，袖口往外翻卷约10厘米。上衣外一般穿深红或深蓝色毛织领褂，亦称"坎肩"。坎肩穿得多者可达3件，肩部从里向外一件比一件窄，有显富之意。下着长裤，裤脚扎彩纹带。脚穿尖头绣花鞋。腰系百褶长形围腰，不用时要按褶纹折叠存放，保持褶纹间平挺整齐。

"七星披"是纳西族最富特色的服饰，因披于妇女背后，亦称"披背"，制作非常讲究，且富文化内涵。

傈僳族服饰

傈僳族分布在怒江傈僳族自治州、迪庆藏族自治州、德宏傣族景颇族自治州、保山市、临沧市、丽江市、大理白族自治州与楚雄彝族自治州等地。

傈僳族服饰（女）

傈僳族按服饰分有"黑傈僳""白傈僳""花傈僳"三个支系，服饰各有特色。盈江一带花傈僳妇女包头帕，帕中段有蓝布，两端用红、黄、白三色布条交错镶拼而成，并钉20厘米左右的小银泡花边。头帕一端饰有红穗。包头时将有红的一端置于左侧，左手按住右手另一端从右向左缠绕三圈后扎紧，然后在扎好的头帕上搭一块红黄白三色拼成的头巾，上绣有箭头纹样，垂彩色绒球和红穗，上衣为蓝布长衫，外罩一件由彩线绣制的坎肩，下身穿长及脚踝的两层围裙，围裙里层较长，下端刺绣有花卉及几何纹，外层较短，长及腹前，用红、黄、白布条镶边，下端镶嵌海贝，绣几何形带纹。小腿上戴藤圈，胸前挂口弦（乐器）。

拉祜族服饰

拉祜族妇女服饰主要有两种类型：长衣长裤和短衣筒裙，其特色在于长衣。长衣为右襟长袖黑布衫，长及脚面，两边齐腰开衩，沿边用彩色条拼合成几何图形，再将银泡排列成相互交错的大小三角形图案，有规则地镶嵌在高领、胸围和袖部位。衣的边缘用波浪形彩带或三角形、长方形图案色布缝合，也有用绣有白波浪图形的条菱形彩带来缝合，再配上精心镶嵌的银饰，给人以光彩悦目、富丽华贵的感觉，下穿黑布长裤，头缠一丈多长的包头。包头有黑、白两种，两端垂有彩色缨穗。

拉祜族服饰（女）

成年人的包头一端垂及后腰部。打绑腿，赤足。男女都有剃光头的习俗，但妇女却在头顶留发一缕，戴大耳环、手镯，胸前佩戴银牌等装饰。

基诺族服饰

基诺族大部分居住在景洪市基诺山，少数在景洪市勐旺乡。

男子外穿开襟无领无扣长袖短衣，以布条拴结为扣。衣长及肚脐处，多用麻布缝制，白色底上织着红、黑、黄色交错的条纹。在衣面饰有一块20厘米的黑底绣花小方块。方块绣有黑边，中间绣红、黄等放射性圆形花纹，周围绣有彩色条纹。这个方块图案，称"月亮花"，是基诺男子服饰的标志。男子喜长布包头。下着裤，戴耳环，耳环洞越大越为美。

基诺族服饰（女）

妇女穿无领长开襟麻布短衣。衣身用条纹布以红、黑、绿、黄以及花布拼镶缝成。背部正中处拼镶一条红色或彩色布，两肩也装饰有一条花边。穿时衣襟敞开，内系一块用彩色布制成的或挑有花纹的挡胸帕。两襟间还系有一块兜形的镶银泡的胸帕，遮住前腹或下胸。襟边无扣，用小布条拴合。裙子短及膝部，独幅，用色布拼成或用有条纹的麻制成。裙腰处镶一块织着条纹的麻布，裙幅两边搭口处镶有花边。膝以下缠黑色布腿套。头戴尖角帽，用五色条纹的麻布对折缝制成。尖角帽像鸟的翅膀，是基诺族姑娘模仿鸟的形状创造出来的。

基诺族妇女喜欢在头上饰鲜花，包头上缀五色缨穗。戴耳环、手镯、项圈，腰佩细篾做成的藤圈。

独龙族服饰

独龙族主要分布在云南怒江州贡山独龙族怒族自治县。

独龙族传统服饰比较简单，但很有特色。男女均袒露臂膀，仅斜披一二张自织的麻布毯。麻毯自一端腋下包抄至另一端肩上拴结系紧。男女小腿部都缠麻布绑腿。未成年的男孩子用藤编一个小笼套挂在生殖器上。成年女子有文面习俗。男子用砍刀断发，腰间左佩利刃，右系篾箩。

20世纪60年代后，独龙族服饰发生了很大变化。妇女已穿内地常见的衣裤衫裙，佩戴耳环挂饰，梳辫顶巾等。男子服饰也多着现代服装，但外衣上加披一方麻毯，以体现民族传统特色。

特殊的服饰传统，造就了特殊的服饰工艺和穿着习俗。独龙族的传统服饰——麻毯，又称"独龙毯"，不仅成为本民族特有的手工艺品，而且成为民族的特色标志。

景颇族服饰

景颇族服饰风格独特。男子服饰以黑、白为主色，老年男子服饰各支系相同，均着黑色对襟短衣和黑色宽管长裤，戴黑色包头。中、青年男子服饰，景颇支系与其他支系间存在细微差别，上身内穿白色立领衬衣、外套黑色圆领外衣，下身穿黑色长裤，头戴红蓝色相间的方格棉纱布圆筒形头巾。其他几个支系的中、青年男子，均着白色衬衣，黑色长裤，戴饰有各色绒球缨穗的白色包头。无论哪个支系的男子出行，均随身背挎筒帕（即背包）和长刀。

妇女的服饰分便装和盛装。若着便装，上身为黑色或各色对襟或右襟紧身短衣，下身为净色或织有景颇族特色图案的棉布长筒裙。盛装是节庆或婚嫁时的着装，上为黑色短襟无领窄袖衫，胸、肩和背部饰有银泡、银牌和银穗，下为用红黑黄绿等各色毛线织出美丽图案的毛质筒裙，腰间系红色腰带，头饰为羊毛织成的红底提花包头，小腿包裹与筒裙质地色泽相同的裹腿，佩戴数串红色项珠及耳饰、手镯。

妇女头戴用织锦对折缝成的筒形帽，上身穿黑色圆领窄短衣，衣肩前后镶缀上百个银泡、银牌和银穗。颈上挂项圈或银链、银响铃等，跳起舞来，嗦嗦作响。耳戴银耳筒，有的比手指还长。手戴刻花银手镯。妇女以戴银多为美。一些妇女还戴红漆或黑漆的藤腰箍。

男子包头巾。头巾一端下垂，上面装饰鲜艳夺目的各色绒球。出门时，腰挂"筒帕"（挎包）和长刀。筒帕为红色，上饰整齐的银泡和银垂片，下有长长的流苏。白刀红包，显出景颇男性的英武之姿。

苗族服饰

苗族分布广、支系多、服饰复杂，头饰与衣、裙都各有特色，异彩纷呈。

（1）昭通花苗。也称大花苗，在昭通、昆明、曲靖、楚雄等地。女子传统的尖髻头饰、花披肩最具特色。尖髻头饰目前改的已多，绣制有传说与历史内容的图案的披肩则普遍保留。

（2）白苗。分布在昭通、文山、红河等地，都称白苗。各地服饰却大不相同。可以分成昭通式、金平式、富宁式与西畴式、麻栗坡式五种。头饰区别最明显，衣服除昭通式外，基本接近，多着开襟短衣、褶裙，多有飘带。昭通白苗也着褶裙，衣为斜襟衣，头盘为薄片裹压圆盘。金平式的头饰为宽包头，一端垂于右胸，另一端垂于额际。富宁式为桶状包头。麻栗坡式，有的把长包头包成叠人字缝。西畴为倒钵状包头，平顶绣八角形纹。西畴白苗还保留绣花传统男长衫。

（3）金平黑苗。以服色为名，包头呈长而圆的桶状，着斜襟衣。

（4）青苗。分布在广南、富宁文山、西畴、麻栗坡等县，各地服饰有差异。富宁青苗盘大黑线辫，小衣长裙，褶裙长至踝，衣对襟紧小。禄丰县也有青苗，戴头箍，着裤，衣前短而后长，前不蔽腹，后下于膝。

（5）红苗。各地有一定差异，盘头、斜襟衣为常见。富宁红苗已改着现代流行服饰。

（6）绿苗。以服色为名，着裙，衣有翻领。

（7）花苗。文山市、丘北县、马关县、麻栗坡县、蒙自市、巍山彝族回族自治县与华宁县都有。文山花苗着披领开襟衣，长围腰、前后飘带、褶裙为主要衣饰，着薄而整齐叠压的圆头盘，有的地方在头盘边缘坠珠串丝须。各地不尽相同。

（8）金平铜厂式。分布在金平苗族瑶族傣族自治县朝阳、铜厂等地。红包头，红腰带，方形围腰，裙与无领圆口斜襟衣为典型服饰。

（9）角辫式。在开远市，主要特点是长而折如双角的线辫头饰。

（10）汉苗。在文山壮族苗族自治州与华宁、金平、师宗、蒙自等县境内，各地服饰不尽相同，金平一带也称其为偏头苗。

（11）清水苗。在金平苗族瑶族傣族自治县。

（12）宁蒗式。在宁彝族自治县，包头呈人字缝叠压，着斜襟长衣，有围腰与裙，也着裤，或于裙内着裤。

第三章　发饰与头饰

云南各民族在装点自身头部的过程中，能创造性发挥艺术才能的部分主要在于发型、发饰、冠饰与耳、面之饰。头虽不大，位置却重要也很能显现装饰效果，所以人们都重视装点头部。头饰的类型极多，艺术性强，设置精巧，在整个服饰搭配中起到极佳的点缀作用。

从历史到现在，云南民族的发型，大的方面可归结为髻、辫、披发与光头几类，各自又有许多细微的区别。发型与发式都体现在女子身上，男性除师宗县的少数彝族留长发外，都遵从现代社会的流行习惯，少有自己的特色。德昂族中，"别列"与"梁"两个支系的女子剃光头，双江拉祜族佤族布朗族傣族自治县也有拉祜族妇女在婚后剃成光头，成为独特的发式。

披发、编辫、打髻三种发式，早在秦汉时期的人物造型中就有，目前仍以这三种发式居多只是形状有差别。

专门做成的固定装饰型披发发式并不多，多为辫与髻的放松休闲状态。披发也就是把松散的长发披在脑后，任其下散，可无其他装饰，也可用头巾之类的东西作简易包裹，也可用绳稍作收紧。

编辫有多种，长发不算短，但也不能满足编制发辫的要求，往往要接上一些东西，常用的办法是接假发辫和接线辫。假发辫是把梳头时落的长发收集起来，编成辫子备用。线辫却是用线编成大辫子，盘在头上作装饰，为了与发辫协调一致，多用黑线，偶尔也有人用彩线，突出其装饰效果。普米族、藏族、哈尼族、苗族、彝族等民族中，都有人使用线辫，纳西族摩梭人也用线辫，彝族蒙古族等一些民族则用发辫。发辫因各地习惯与婚姻情况差异，有单、双辫与多辫等多种不同的编法，盘发辫的方式也有差别。

发辫要用头绳扎紧，多数还用纱帕、头巾与头帕包好发辫，并盘成一定的发型，露发于外的并不多。

云南历史上最有名的发髻是椎髻，各个时期，人们所指的椎髻不尽相同。

秦汉时指的椎髻，髻成纺锤形，中腰扎紧，两头大，拖扎后颈部位。后世所谓椎髻，多数都是指盘髻头顶而言。发髻都包在各种头饰物之内，很少外露，平时难得一见，也不是很重要，多数是为了造就头型，作简易打髻，相对较为随便，但也有许多固定不变的发髻。常见的有顶髻、拖髻、散盘、钵形髻等发髻。顶髻指盘头而言，有螺髻、尖髻、莲蓬形髻、柱状髻等多种。盘发随便，命名也不能十分准确。

装饰在头发上的饰品有簪、钗、发夹、饰片等多种，有的用于固定发型，有的起装饰作用。

簪是使用较普遍的固定发型用具，也起装饰作用，早在秦汉时代，云南的各族就已大量使用。簪身为一长尖状体，插入发髻中起固定作用，主要花式出现在簪的尾部，多制成花鸟虫鱼形象，装饰有的制成半球形，镂空刻花，也有的刻花成条状，还有的制成鸟、蝶等造型，式样极多。用材除金、银外，也用玉、骨牙、竹、木等。

钗的作用类似于簪，钗为双股，也是在尾部作图刻花装饰。有的还垂挂链饰挂件作装饰。通常深入发中，钗可高出发饰，加上挂饰，有独到的装饰效果。瑶族的头钗就是如此，插入发髻之中，外露钗尾与挂饰。

在头顶加入特制的竹筒，与发盘结，外包头定型，是西双版纳一些哈尼族的饰头方法，其竹筒为独特的饰物。

瑶族使用银制太阳纹圆顶盘，也是圆形头盘，其中银条连接中间为骨架，也有的为了方便，用柊叶、芭蕉叶、笋壳等制作。

勒子是许多民族都有的饰物，装饰在前额上部发际，为弯曲半圆条，多刻花纹图案，或饰乳钉，还有的坠链坠作装饰，也有的用布或其他物品制成，缀珠、玉、丝须等物作装饰，也称作头条。

装饰梳也是头发饰物，有银制品，也有其他制品，有的直接插入发中，也有的是与链坠连在一起。头箍以佤族用者为著名，既作装饰，也用于固定发型。红头瑶的银头箍则属另一种类型。

编辫或打髻，都要用纱巾、丝网、头巾、头帕包和帽子之类的东西包住头发。过去用黑纱巾和丝网巾包住头发，再包头巾或戴帽，也可以只包纱巾。目前流行的花头巾都是机织方巾。

包头、头帕大同小异，不能以布的形状来区分，多数都是长方布，也有短帕，包成一定形状固定套用的，也称作头套。以包好置于头部时所表现出的形状差异，云南各民族的包头可以归为30种左右。

（1）高桶状包头。圆如柱，高超过直径，状如桶，也可能上部略小于下部，有的还在外围饰璎缀等物，有的用布包裹，也有的是固形套上。元谋县凉山支系彝族、师宗等地彝族、元阳县壮族、景颇族、花腰傣、怒江傈僳族男装，都有这种包头。

（2）平桶状包头。也成圆柱状，直径大于高度，外缀缨须，金平、河口等地红头瑶有这种包头，富宁县彝族中的花倮都有这种包头，厚薄不一。

（3）大轮盘包头。小布带层层叠压，裹如轮盘，留下中空部位套在头上，厚不过几厘米，直径则有大有小，可分成数种，边缘也饰丝须、串珠之类的细碎饰物。文山等地苗族、大姚彝族、新平哈尼族卡多人，都有这种头饰，卡多人还以交叉布带勒住头盘，带上饰银泡。

（4）绞裹式包头。男女都用，整体形状为圆形，中空纳头，裹时不要求整齐，形成绞缠，禄劝武定等地彝族还把这类头饰裹好定型，用时套上，四边饰花、蝶等形银饰件。

（5）人字缝包头。主要特征是使前头部位的包头，叠缝折成层层叠压的人字形，苗族、傈僳族等民族都有这种包法。

（6）短帕包头。类型较多，有的用短方布，也有的用毛巾包头，有时在头帕两端作图案、绣花，或缀璎须、珠子，包时外露作装饰。毛巾包头多配便装，许多地方都用。

（7）头套。外围用硬布制成圆套，中空部分以布覆盖，也可让头发外露，各随其俗。石林、泸西、弥勒等地的彝族撒尼人，石屏县的哈尼族就有这样的头套，因厚薄、长短与花色差异，形成不同的类型。

（8）翘角包头。在包头时，有意留下上翘的包头尾端，形成独特的头饰，壮族侬支系、布朗族为典型。

（9）伏瓦状包头。使包头外观如两板相搭，如瓦屋顶马关县的傣族，石屏、元江等地傣族，金平苗族瑶族傣族自治县的瑶族沙瑶支系，都有这种头饰，有的地方是未婚女子的装束。

（10）倒置背箩状包头。包头如扁圆柱状，上部略小，如背箩，石屏、元江等地傣族有这种包头。

（11）倒钵状包头。形如倒置之钵罩在头上，浑圆而上大下小，苗族与景谷等地傣族有这类头饰。

（12）留角包头。在包头的前端或其他部分留出尖角，有一角，也有多角。元江哈尼族彝族傣族自治县羊岔街的哈尼族、保山傈僳族、石林彝族自治县的彝族白彝，都有这种头饰。

（13）方巾包头。方巾为新式机织产品，20世纪80年代及以前曾经在很多地方流行，多包在辫与髻或其他包头之外。

（14）大理白族式头饰。主要流行于大理市、洱源县等地，丽江纳西族自治县和其他一些地方的白族，也有此头饰。

（15）蓝靛瑶头饰。太阳纹圆顶盘，饰链坠与相接柱状发髻是主要组成。

（16）西畴花倮包头。流行于西畴县鸡街乡及邻近一些地区。

（17）小凉山彝族式头盘。婚前婚后不同，片状盘是主要标志。

（18）搭折包头。壮族土僚中的搭头土僚、屏边彝族、巍山西山彝族都有这种包法，布条搭折包头是主要特征。

（19）尖状包头。头饰尖突，有的盖布，也有的裹布，还有的有大量银饰品，并有后饰长带。壮族中的尖头土僚、黑沙人，哈尼族中的尖头阿卡，金平、河口的红头瑶都有这类包头，相互间的外观差异非常大。红头瑶属婚后妇女装束。

（20）长衣包头。主要流行于澜沧拉祜族自治县及西双版纳个别地方，包头就是上衣，是哈尼族的一种特殊头饰。

（21）南美拉祜包头。主要在临沧市南美乡一带。

（22）彝族花腰包头。也称作帽子，在石屏县与峨山彝族自治县，制作复杂，穿戴也复杂，戴好后，后面的花鸟图案极醒目，装饰效果极好。

（23）垂带式包头。包头包好后，两端从内侧下部露出并下垂，成为装饰。

（24）大黑彝式包头。流行于石林、泸西、弥勒等地大黑彝中，是一种特有的包法。

（25）包头加盖式。在包头之上盖布作装饰，彝族、哈尼族、基诺族等民族有这类头饰。

（26）前平后斜式包头。壮族平头土僚、彝族中的姆基人有这种包头。

以上只是一些常见的包头法，包头头式极多，变化也有一定的随意性，很难列全。

包头之外，帽子也是头部重要遮盖物和装饰物。帽子男女老少都有，儿童的帽子形状、式样为最多。有的童帽近似于大人的帽子，更多的是独具特色。金鱼形帽、虎头形帽、狮头形帽、孔雀帽、猫耳帽、花鸟形帽、喜鹊帽、牌坊帽、鲸鱼头形帽、圆罩形帽、兔耳帽、莲花帽、水仙帽等许多童帽都很常见，不限于单独一个民族。西双版纳哈尼族童帽、瑶族童帽、彝族鸡冠帽，新平、元江等地傣族童帽等，则是有民族特色的独有帽子。

成人男子帽以瑶族的马尾帽、纳西族东巴五佛冠、景颇族脑双帽、藏族皮帽、回族白帽最具特色。

女帽则以基诺族、孟连富岩佤族与哈尼族奕车人的尖顶帽，大姚县桂花彝族的罗锅形帽，鹤庆县彝族白衣人的罗锅形帽，屏边苗族自治县彝族孔雀开屏状帽，福贡县傈僳族族串珠网帽，元谋、武定等地彝族红彝支系花帽、鸡冠帽，新平彝族傣族自治县戛洒傣族少女帽，禄劝、武定彝族毛线帽等最有代表性。

鸡冠帽是彝族中流传极广的一种帽子，大致可分为红河式、元阳式、绿春式、阿乌式、蒙自式、楚雄式、禄丰红彝式、昆明西山式等7种。楚雄式也称作鹦鹉帽，禄丰红彝式也称作蝴蝶帽，通海县彝族的喜鹊帽也是一种变形，巍山彝族回族自治县西山彝族未婚少女也有鸡冠帽。此外，白族也有类似西山式的鸡冠帽，称为凤凰帽。红河县乐育乡一带的哈尼族有与彝族相同的鸡冠帽。

现代的檐帽、军帽等，在许多地方也很流行，男女都有人戴。军帽由于布料限制，出现时少见，在20世纪六七十年代为稀罕之物，许多以得一顶为荣，从此流行，沿至今日。

在帽子与包头之上，往往还要有点缀，帽子多饰雕神佛像和寿、喜等字的银牌，还坠有鱼龙等饰件。包头的饰件还更多，有银牌、银链等外，还缀有缨球、丝须等物作装饰。玉石、玛瑙、兽爪、兽牙、海贝、料珠等，也是常用的头部饰物，锦鸡、野鸡等长尾鸟的尾羽也用于头部装饰。

耳饰也是头饰中的重要内容，云南各民族中常用的耳饰有耳珰、耳环与耳坠，耳珰也称作耳柱，直接塞入耳垂上的穿孔中作装饰。比较常见的耳珰有圆筒状耳柱、蘑菇形耳柱、圆筒饰菊、花形头耳柱、莲花形银包玉耳柱、长杆形饰缨须链坠耳柱等，耳柱使用的民族不多，主要在德昂族、傣族、布朗族、佤族等民族中使用。

耳环与耳坠的使用很普遍，几乎每个民族都有。耳环的造型很多，通常可以成细圈金银耳环单股头尾雕花刻图耳环，环身饰刻花金、银片耳环、环身饰玉片耳环等。单股头尾刻花制图者，通常制成弹簧形、蕨芽形与花形等。耳环上的刻花图案多以花鸟虫鱼草木之类为多。

耳坠多由耳环与坠子两部分组成，环多为金银制的线圈，也有较为复杂的造型。坠子部分比较复杂，有制成叶、果花等形状，也可以制成其他自然物与人工制品造型，还有大量的部分是挂链坠，也有动物造型。

第四章　肢　饰

　　肢饰主要有四肢部位的衣装花纹图案与饰物佩件。花纹图案主要出现在上衣的袖子与裤管、裙边、绑腿、鞋之上，既讲求全身协调，也各有其特点。饰物则有指饰、腕饰、臂饰、足饰、踝饰与腿饰。

　　裙布垂至下肢部位的部分，是花纹与图案的主要载负部位，通常用接布拼色、打褶、织花、染图等办法来制成图案花色。蜡染者有苗族与麻栗坡新寨的彝族，织花以景颇族为精美，打褶的民族则很多。也有的民族在裙摆边缘绣花边作装饰。

　　裤管的花纹图案装饰可分为几种，有的在裤管中缝配红、黄、白诸色线作装饰，有的在裤管边缘缝上花边或绣花作装饰，也有的在整件裤管上绣几何纹或花鸟图案。裤管边缘装饰花边，过去许多地方的妇女服饰都有，绣花则见于拉祜族等少数几个民族，有一种拉祜族图是绣成太阳纹。绣几何纹以金平、勐腊等地的瑶族为典型，绣有树枝形组图、万（卍）字形组图与其他多种几何纹拼图，并以红、黄、蓝、紫等配色协调各种图案，造成醒目的主色调。全裤绣花的服饰也有多种，以永仁县彝族服饰为例，构图素材为花鸟，采用花边与绣图配合、素色与彩色搭配的办法，绣出精美协调的裤子装饰。富宁县的彝族则以三角形拼花构图，并以红、黑、黄、白等色调搭配做成美丽的图案。

　　绑腿有长布带、斜角布片与套筒三种。长布带的装饰以边缘缀花穗为主，也可在带边饰线条，绑时绕成花线条，缀花穗以石屏县龙武、哨冲一带的彝族为典型，当地称这类饰物为杨梅花，多是男子用。斜角布片与套筒都可以绣花，但绣者有之，不绣者也有之，各随所好。配制精美的绑腿在小腿部位也有极好装饰效果，往往有裙下着裤或大管裤下藏小裤的感觉，有的装束膝部露出，上有裤、裙，下有绑腿，也有独到的装饰效果。

　　袖子构图有大小袖套穿、单袖接布套色、花色线圈装饰、前臂绣花、袖口绣花、中袖绣花与全绣花几种。有的服装，袖子与衣服是分开的，着装时各自套上固定，才成一体，昭通苗族的独立大袖就是一例。大小袖套穿主要出现在过去流行的姊妹装与半截观音一类的衣服上，这类衣服的袖子，大袖短，小袖则长，上下搭配，就同大袖中接出小袖一般，各自都绣花，有独到的装饰效果。壮族、布依族、彝族、蒙古族等一些民族中，目前仍有人保留这种衣服。

接布套色的袖子，有的用相同布料，不同颜色，造成效果，也有的用土布接锦缎的办法造成特殊的装饰效果，锦、缎有花，就接上了一条花袖套。哈尼族、壮族、彝族等许多民族都用此法。

线圈装饰主要在袖子上装上红、黄、白、蓝等不同色彩的线条，形成线圈，以色彩搭配突出装饰效果。基诺族、拉祜族等都用这种办法装饰袖子，有时也用小布条制成线圈。

不论是前臂绣花、中袖绣花、袖口绣花还是全袖绣花，题材多是蝴蝶花鸟一类的图案，袖口绣花多绣成半圆或球面三角形，以蝴蝶、牡丹一类的花鸟图为多。前臂与整袖图案多成圈套接，从上到下层层叠图，有时还会杂花边作装饰。麻栗坡县新寨一带彝族的蜡染袖子图案与昭通等地苗族的几何纹粗线图案是比较特殊的两种。还有的在袖口加花布或黄白红等线圈作装饰。

鞋又分为鞋和靴。靴即有筒鞋，以藏族皮靴使用最多，有的地方也保存布制靴子，有筒绣花，目前已很少有人做。鞋的类型很多，以材料分，有草鞋、布鞋等。布鞋又有一般的浅底鞋与布凉鞋之分。哈尼族及其他一些民族使用的木屐、树皮鞋、竹草鞋，则是用料与形制者比较特殊的生活用品。

布鞋的形制，除帮与底的造型外，鞋帮花纹、鞋头图案是区别之所在。以鞋头图案分，有斑鞋、猫头鞋、鱼形鞋、十二生肖鞋等多种。以鞋底鞋身形状论，则以船形鞋为特殊。鞋帮绣花以牡丹蝴蝶与花鸟虫鱼等图案为多，还有老鼠葡萄仙桃佛手一类有吉祥含义的组合图案。

指饰品主要是戒指、指环与指箍几种，指环多是简单的圆环，用金、银、玉等材料制成。指套有的是指环的扩展，把金、银制的指环连线叠压如弹簧圈，就可称为指箍。还有一些是圆环套筒，可以饰花，也可以素用，有些指套造型类似大臂箍，只是缩小以适合套在指上。

戒指的造型相当多，往往随心所意都能做，包括指环与戒面两部位，主要在戒面上做文章。戒面有圆形、方形和其他多种形状，以方、圆二形为多，也有盾形、六角形、动物或其他几何纹造型。戒面有素面、刻字、币面、刻花等多种构图。刻字者可刻单字，也有的制成印章。刻花者有单刻花鸟的，也可刻上喜鹊登梅一类喜庆组图。还有以镶珠宝玉石作装饰的，有单镶，也可镶多颗。有的戒面还镂空制图，有的戒指还饰小链坠。

　　腕饰主要有手镯、手链与手箍。手镯有玉、翡翠、琥珀与金、银、铜、铁等不同的质地，有草编镯，为儿童用物。玉类手镯多制成圆体光身，不刻花纹图案，以其本身的色泽质料取胜，偶尔也有包银片金片的。铁镯只有独龙江等极少数地方偶尔可见，数量不多。铜镯因价贱，加工也不太讲究，以圆环扭丝镯等为多见。

　　金镯，银镯的加工则要复杂得多，有镂空雕花、绞丝、刻花、塑形、雌花等多种办法。常见的手镯镯身形状有四方形薄圈，圆形、六棱形、鼓形、扁圆形、扭丝、绕丝叠圈等。有的镯做成圆环状，从手指部位套入，也有的有开口，从腕部卡入。镯身图案多为花鸟虫鱼、瑞兽神龙。刻花图案多刻花、蝶、树、叶之类，也有犬牙、回纹及其他一些几何纹。堆花饰有狮子、麒麟等瑞兽纹，镯头对口则有龙头、龙戏珠、龙衔珠等造型。有的手镯饰链坠作装饰。

　　镯的形状，不仅各地有差别，男女老少不同性别年龄也有差别，主要体现在大小厚薄与花纹图案上。佩戴法也有差异，少者戴一只，多者可同时戴五六只或更多，自腕延至小臂。传统的手链多属儿童用，链上有长命锁，既作装饰，也起护身符的作用。

　　手臂部位的饰物有臂钏与臂箍，臂钏与镯接近，只是圈更大。臂箍多做成长筒或短筒状，卷口，箍身刻花或镶花饰漆彩作点缀，也有筒状开口的，还有弹簧卷状的。长臂箍的长度达9厘米左右，或更长，臂箍用的民族不算太多，有佤族、景颇族、哈尼族等民族，用于小手臂，连腕部的臂箍，也有人归为手镯。

　　踝饰以脚链为主，属小儿饰物，缀锁，作护身装饰之用。佤族等一些民族有竹与藤制的腿箍。

第五章　附　件

附件是一些独立个体，多数都具有一定的功能，是工具，同时又成为某个民族或一些人固定使用的随身物品，起到装饰的作用，是服饰系统中的特殊成员，也有一些是纯粹的装饰品。

云南民族民间常见的附件有香包、挂件、挂盒、槟榔袋、绣球、护身符、围腰、背被、箩、草帽、篾帽、伞、蓑衣、挎包、钱袋、搭链、刀、弩等。

大理地区的白族、彝族都有香包，内装香料或五谷等物，既可悬挂作装饰，也可佩戴，过去认为有避邪之功。常见的形状有人、猴、虎、蝴蝶等动物，也有佛手、石榴、花篮等形状，包身还绣有各种吉祥图案和喜寿一类的吉祥字。还有小挂件，用途也差不多。

有些小绣球，也同香包一样，可以佩戴，壮族、白族、傣族等一些民族有绣球，做工多较精致。

槟榔袋是文山壮族苗族自治州一些地区的瑶族使用的装饰小袋，用丝线编长带挂在身上作装饰，袋下缀丝须，过去用来装槟榔，故有此名，现已成为一般性装饰品。

挂盒则是藏族使用的一种大型装饰品，也有护身作用，佛龛形，方形饰金刚杵等造型，就是其中的几种。

竹编箩在一些地方，既作载物的工具，也是随身物品。出门往往就背在身上，即使是空箩也是如此。这类箩有大有小，小的以元江、新平等地傣族的腰箩为典型。腰箩傣语称作秧杆，拴在腰间，装小件物品，又是装饰品，年轻人用的腰箩、箩身还饰有线球、线须等物，极为精美。

有的民族有绣制精美的钱袋，扎在腰间，既放钱物，又是饰品，布依族等一些民族有保留。白族制作的腰掌，有点类似于钱袋，也围在腰间，只是没有放物品的袋子，有如兜肚一般。

草帽与篾帽是防晒避雨的工具，但在很多地方，也在草帽或篾帽上装上绣制精巧的带子或银链，平时出门即带，用作随身饰物。新平元江等地傣族的两种小篾帽，则反以装饰作用为主。其一类似盛开的鸡枞，另一类似才开的鸡枞，顶在头上作装点，有的还上漆彩。

新平、元江等地的傣族，不戴篾帽，以伞作装饰，出门即带。壮、瑶、彝等民族也有类似的习俗，还制作精美的伞套，平时背在身上做饰物。

蓑衣是雨具，也是服饰家庭中的特殊成员。云南民族民间做的蓑衣有几种，用山草、棕、竹、露兜树等材料制成。材料不同，衣的形状也不同，山草编结缝好，较为费时，竹破丝连接而成层层叠压，使雨下滴而不能渗入。棕取其叶柄上的纤维网制成，露兜树则取其长叶片去边刺晒干，缝合成雨披。各有长处，也有不足。

背被用于载负小儿，又是精美的工艺品。背被因为宽大，适于绣制图案，多数地方都要在背被的中央与四周绣花作装饰。就云南全境而言，文山壮族苗族自治州的壮族，大理白族自治州的白族，楚雄彝族自治州、巍山彝族回族自治县、禄劝彝族苗族自治县、石屏县、红河县、石林彝族自治县等地的彝族为绣制工艺出众者，苗族的背被也有精美的绣花。这些背被不仅绣工好，图案的文化内涵也极其丰富，有如花鸟画的殿堂。以武定县彝族的一件背被为例，绣有花、石榴、鸟、蝶桃、鱼戏莲、兔子拔萝卜蟹佛手、老鼠葡萄、喜鹊登梅、菊花等，背被上缘还绣松、竹、梅，梅上有喜鹊，题字云梅花开富贵，喜鹊报平安。壮族背被的内容同样多，群龙舞云、花鸟鱼蝶等构图，有千种万种，每一个小图都有文化内涵，小图又可组成大图。白族的背被绣花，突出大红色，绣制紧凑严密，来自传统的文化内涵也很多。

各地不同的背被外形也有极高的艺术价值。刀是云南许多民族不离身佩物，可防身，做工具，也是工艺品，刀分长刀与小刀两类，小刀的佩带，藏族、彝族、阿昌族、回族较有代表性，彝族带刀者以青年人为多。长刀则以阿昌族、景颇族、怒族、独龙族等民族为代表，怒江傈僳族自治州的傈僳族、怒族、独龙族三个民族，标志性长刀都是只有半边刀壳的挎刀，一边的刀身外露。景颇族与阿昌族用的刀都来自阿昌族聚居区，外形接近，区别在于刀身，阿昌刀有刀尖，景颇刀为齐头。

弩与箭包在怒江地区也成为重要的随身物品，不仅上山，赶集也有人随身佩带。挎包每个民族都有，也是重要的饰物。景颇族、基诺族、佤族、德昂族、傣族、西双版纳哈尼族等，包的饰物与色调与服装相同，地方特色与民族特色都突出。景颇族的挎包色调同筒裙，装饰的银泡与坠子则类似于衣服，只是小一些。基诺族麻织挎包的装饰线，一如衣服裤子。佤族挎包的色调同裙子，以薏仁装饰。德昂族挎包则如同衣装构图，布料色调与绒球等装饰都与衣服接近。西双版纳哈尼族的挎包，一边有图案，绣几何纹等图案，一如衣服上的图案。彝族的包，不同于服饰，品种极多，工艺特别。

西双版纳哈尼族、布朗族、文山壮族的葛藤纤维包和网袋与瑶族、傈僳族、彝族等的麻纤维网袋，也是一种特殊的挎包。

大理白族的绣花包，色身为硬壳，绣花风格统一，形成自己的特色。

第六章　服饰工艺

扎染、蜡染工艺

进入大理地界，在引人入胜的美景中，首先映入眼帘的除了高耸入云、终年戴着雪冠的巍峨苍山和蓝莹清澈的高原镜湖洱海外，最突出的要算以蓝白花为基调的手工特产扎染布了。

扎染布是白族特有的工艺产品。在大理城乡的街道店铺、摊棚、旅舍、宾馆、酒店以至居家住宅内外，随处可见它的踪影。它或者被人做成衣服、裤子、裙子、马甲、领褂、凉帽、便鞋、手绢、头巾、围腰、手袋、挂包、背包、腰包，或者被加工成床单、被套、帐围、桌布、枕帕、窗帘、门帘、壁挂，电脑防尘罩、电视机罩，或者整幅整匹地陈列着，供人欣赏选购。走到哪里，端详一下白族妇女的衣着和头饰，也留意一下周围其他民族身上穿的或佩挂的饰件，差不多都能寻到扎染的影子。扎染不仅代表着一种传统，而且已成为一种时尚。当旅游者从当地把扎染天南海北地带到全国各地甚至国外时，它从远古时期走来的脚步仿佛与当代步伐连接到了一起，叫人不能不格外惊叹于它所具有的与众不同的魅力。

扎染如此受欢迎，是由于它特有的有别于其他染织物的个性。它朴素自然，蓝底上的朵朵白花清清雅雅，毫不张扬，符合人的情致，贴近人的生活，充满人性色彩，是白族人民勤劳、质朴、纯洁、诚实、善良和乐观、开朗、热情好客等美好品格和情趣的体现。传说它是苍山的溪水所变，或说是仙女织出带给人间……穿用它不仅美观，还代表着灵巧和智慧，更能体现永恒、亲切、真诚。几乎无所不在的扎染，在人们心目中差不多已成为大理最耀眼、最特殊的文化象征和民族传统艺术的标识。

扎染如此，和扎染基调相似的苗族蜡染也一样，它像一簇簇烂漫的山花开遍云南全省苗族山乡。作为旅游商品和文化特产，它早已进入城市并且越来越多地成为人们喜爱的重要民间艺术品。它的白白蓝蓝的情调宛若赞美着苗乡，歌颂着山水自然，展示着苗族人民的性格和美德，散发着永恒的泥土芳香。

扎染。

扎染古称绞缬，是由来已久的一种染布工艺。大理叫它为疙瘩花布、疙瘩花。疙瘩花布也叫作结花布、蓝花布，扎染是近些年来才逐渐叫热的名称，它是一种先"扎结"或"扎缝"而后漫染的蓝白色相间为主的染花布。因主产地在大理，染布者绝大多数是白族，故而人们又习惯把它叫作大理扎染，白族扎染（或疙瘩花布），是古今驰名、备受人们喜爱的手工扎染产品。

扎染工艺出现较早，宋代《大理国画卷》所绘跟随国王礼佛的文臣武将中有两位武士头上戴的布冠套，同传统蓝底小团白花扎染十分相似，可能是大理扎染近千年前用于服饰的直观记录，大理地区明清时期的寺庙，曾发现有的菩萨塑像身衣有扎染残片，有扎染经书包帕等，到了民国时期，居家扎染已十分普遍，以一家一户为主的扎染作坊密集著称的周城、喜洲等乡镇，已经成为名传四方的扎染中心，方圆百里的许多白族村民除前往购买扎染以供衣饰之用外，还去那里学习技艺，然后回家自扎自染，由于这种习惯，所以扎染工艺也就不胫而走，传播面越来越宽，会扎、会染的人很多。

中华人民共和国成立以来，随着经济发展和人民生活的改善，扎染的需求量逐年增加，扎染的产量和花色品种与日俱增。特别近二十年来，在改革开放的推动下，在蓬勃兴旺的旅游业需要大量旅游产品的形势刺激下，扎染出现了前所未有的兴旺景象。扎染已跃出单家独户的生产方式，出现了一些集体作坊和具备一些新技术设备的扎染厂。

扎染用的布料过去完全采用白族自己手工织的质地较粗的白棉土布，现在土布已较少，主要用工业机织生白布、包装布等布料。布料须吸水性强，加工主要有扎花、浸染、漂晾三道工序。扎花，原名扎疙瘩，即在布料选好后，按花纹图案要求，在布料上分别使用撮皱、折叠、翻卷、挤揪等方法，使之成为一定形状，然后用针线一针一线地缝合或缠扎，将其扎紧缝严，让布料变成一串串"疙瘩"。接着便是浸染，即将扎好"疙瘩"的布料放入染缸中浸泡，经一定时间后捞出晾干，然后再将布料放入染缸浸染，如此反复数次，最后捞出放入清水将多余的染料漂除，晾干后将线拆除，将"疙瘩"挑开，被线扎缠缝合的部分未受色，呈现出空心状的白布色，便是"花"；其余部分成深蓝色，即是"地"，至此，一块漂亮的扎染布就完成了。"花"和"地"是由于受色与不受色两种浸染结果造成的，两者之间往往还呈现一定的过渡性渐变的效果，即蓝色底与白色花之间不是生硬地形成对比，花的边缘都有渍印造成的渐淡或渐浓的色晕，使得花色更显丰富自然。许多时候，这种渐变造成的渍印效果是通过掌握扎缝疙瘩的松紧来人为地控制的。传统染料以植物染料为主，从前主要用板蓝根，由于其提炼的染料上色慢，浸染工效不高，植者收益甚微，加之种者极少，因而不少人大量改用蓝靛浸染。现在，使用化学染料者正在增加，古老的浸染方法面临着逐渐消失的危险。然而，坚持传统染法即用植物染料浸染的传统仍在一些人家中保持着，许多供自用的扎染也常以植物染料染之。这是因为人们依经验和习惯认为，板蓝根或蓼蓝叶等植物染料与其他的化学染料相比益处颇多，植物染料最重要的是色泽自然，褪变较慢，不伤布料，经久耐用。化学染料染出来的布，开始时看起来色调沉艳，但是经过反复洗涤和日光曝晒，就会越来越淡并逐渐失去自然的色感。植物染料则完全不同，经反复浸染的布料，随着时间的推移，花与底的色彩反而会变得越来越明丽、清晰、谐和，用它缝制的衣裙，色彩鲜活、明快，穿着时还能带来舒适感，不会对人体皮肤产生不良刺激。据说，像板蓝根一类的染料同时还带有一定的药物作用，对人的健康有益。大理周城是白族扎染最集中、产量和品种最多的地方。那里的白族妇女人人会扎花，她们利用一切可以利用的时间为自己或从别家领件来不停地做扎缝"疙瘩"和浸染工作。田间地头，或摆摊开店，甚至坐车外出，箩筐里总是背着布料，随时进行扎花之事。她们同男子一起进行浸染工作，出产的扎染工艺品丰富多彩，深受群众喜爱。

扎染的花纹图案琳琅满目，选材上主要包括植物花纹（如花、草、叶、果等），动物花纹（如虫、鸟、兽、鱼等），自然花纹（如日、月、星、云、石、山、水等），人物花纹和吉祥纹等五类。其中植物纹运用最多的是植物花果纹，居首者是梅花，其次是浮萍花（碎点花）、小团菊花、茶花、素馨花、刺叶花、豆叶花、桃花、菱花等，花纹有的含有一定意义，有的纯粹为了表现自然物的形态美，有的则是借物寓情，通过对花的描绘来反映人物的心理状态和思想意识。动物纹中用得最多最常见的是蝴蝶纹（即蝴蝶花）。蝴蝶是母亲和风调雨顺的象征，它的重要性和使用的普遍性使它具有扎染纹饰母题的地位，大蝴蝶、小蝴蝶、单蝴蝶、双蝴蝶……各式各样，在扎染中举目可见。现在，扎染布的花纹有了许多新变化，除传统纹样仍大量存在外，由于商业化影响，有的厂家为迎合订货者的喜好，按对方提供的纹样设计制作了一些产品，使非当地传统的国内国外纹样不断出现在扎染商品中。

蜡染。

蜡染工艺和扎染工艺从外观上看好比一对美丽、纯洁的姐妹。二者在制作工艺上有相似的地方，也有不同的地方，可谓异曲同工、异彩同辉。

云南蜡染工艺以苗族蜡染最有代表性。云南苗族大多居住于崇山峻岭之中，靠山地种植、狩猎和饲养度日，自给自足的自然经济占了重要地位，纺织业在苗族家庭中是主要的副业，是他们衣服用料的主要来源。苗族妇女自幼便跟随母亲学习纺线织布，长大后多数都是纺织能手，因而纺织业成为社会分工中完全由妇女担当的事。这种状况在今天多数苗族地区仍然保持着。一些古老的纺织手法，像腰机织布、活动木架机织布等在苗乡仍普遍可见。她们的纺织产品以白麻布为主，棉布次之。古老纺织方法织出的布，正是蜡染需要的理想布料，它那表面粗糙、质地铁实的特殊风味，是别的布料难以代替的。这大概是苗家传统织布一直保持着的一个重要原因。

苗族蜡染同白族扎染在制作原理上是一样的，只是方法各有不同。蜡染一般有以下几个环节：一是准备好布料。苗族蜡染所用的布料是自织白麻布。苗族种麻，各村各户都有麻地。每至麻成熟收割后，人们便将麻皮撕成细丝，绕成团，然后拿到木纺车上绩纺，将麻纺成线后便用自制的漂白剂进行漂白（一般是用石灰或草木灰煮和浸泡），漂白后的麻线因质地粗糙不易织制，因而还要用木滚筒在石板上用脚驱动着将麻线碾光碾滑，之后再将麻线投入掺有黄蜡的清水中沸煮一次使其变得柔软，最后用回线车将麻线退为麻团，即可用腰机织成白布了。布料有了，接下来的第二步是最重要的——点蜡花，即用蜡在麻布上画出要染的花纹。点蜡花的方法比较古朴，具体做法是用蜂蜡与树脂混合，放于一个小土碗中，置于火塘上加热待蜡熔化后便用专制的蜡刀（铜质，刀宽约2厘米，形似小斧，连接在约15厘米长的竹柄上）蘸蜡往布上涂绘图案纹样。涂绘时蜡的厚薄要均匀、适度，为此蜡要一直置于火塘上保持温度，保证蜡的稀释度符合涂绘的要求。蜡花画好后，让其平放或挂起来自然风固化，经检查和对残缺处进行补蜡后即可拿去浸染。之后进入第三个环节——浸染。苗族浸染原料同白族相似，都是采用植物染料，而靛蓝是他们最常用的一种染料。靛蓝的制作过程是把新鲜的蓼蓝割来，置于清水中浸泡，直至泡烂，然后取出它的秆和叶，加入少量的石灰水再泡，使其发酵，发酵后的靛汁即是染料。染布时将一定量的靛蓝放入水池中稀释溶解，然后将点好蜡的麻布放入其内浸染，不停地用手在染缸中搅动，使布均匀受色。经一定时间后，将布捞出晾干，又再次放入染液中浸染，如此反复几次，最后将布料取出放入煮沸的清水中，使蜡脂脱去，复用清水揉搓漂洗，除尽蜡屑，晒干，在浸染过程中，涂绘蜡花的部分不会受色成为白花，未涂绘腊花的部分则会受色，成为蓝地或蓝花，这样，一块蓝白相间的美丽蜡染就制成了。

苗族蜡染工艺产品多用于制作衣服、裙子，特别是女子穿的百褶裙。许多女子服饰以蜡染布为地，加上精美鲜艳的刺绣使其更锦上添花。其次用于做床单、被面、头巾、挎包等用品。由于蜡染具有独特的实用性和审美价值，它已走出苗家山寨，大量进入城市人的生活和社会艺术领域，已较广泛地被用来制作各种新式的帽、衣、裙、背包、挎袋、桌布和墙上装饰如壁挂等，其中蜡染绘画已成为一种备受青睐的艺术新品种。

云南苗族蜡染清新活泼、粗犷大方、纹样丰富、图案多变、构图严密、形象生动、造型独到，具有浓厚的乡土气息和民族特色。蜡染图案纹样中几何图案最为常见，其次是自然物形图纹如蕨草纹、豆花纹、睡狗纹等，纹样造型独具一格。各地苗族对蜡染花纹的喜好不一。滇西南苗族（花苗和青苗）蜡染的几何纹较多，滇东北苗族蜡染中自然物花纹占有较多分量。以几何纹、齿纹、菱形纹、十字交叉纹、弯转纹、带状纹等为主的蜡染，基调素雅、线条清晰、设计规整，整体构图简洁大方。连续反复的花纹给人一种连绵不断、美不胜收的感觉，带有强烈的艺术魅力。以自然物纹为主的蜡染，色泽清雅，着力表现出丰富、自然、和谐的美感，多变的纹样幻化出生动活泼的个性，充分反映了苗族人民对大自然的热爱之情和由衷的赞美。把蜡染比作散发着泥土气息、生活味和山地民族纯情的诗画，是恰如其分的。

扎染、蜡染的艺术特色

谈到扎染、蜡染的艺术性首先要说的自然是它那令人难忘的蓝色及由此产生的蓝色情调。蓝色是一种富有深刻内涵和浓厚感情意蕴的颜色，是一种富有节奏韵律的颜色，是一种能够呼唤浪漫联想的颜色。它清新、朴雅、恒稳、沉静，就像一座连接事物与人的感情的桥，能将人的思路引向宽广的原野，让人从充满生命力的种种物象中找到人与自然、人与人和谐、亲近的感觉。乍看它似乎有点儿粗朴单调，仔细端详品味，却包含着丰富的变化，给人以无尽的遐想与喜悦。人们看着、拿着或穿着扎染或蜡染的时候，看到这蓝色，立即会与蓝天、海水、秋山、茂林等联系到一起。它像蓝色写成的诗话，装点、美化着人们的生活，让人们从它的基调中获得艺术美感。

扎染、蜡染都是云南少数民族的一门古老手工艺术，它们在色彩主调和工艺原理上相同，而在艺术风格的具体表现上存在着一些差异，各具特色。扎染花纹取决于缝扎疙瘩的情况，手上功夫是关键，由于扎缝疙瘩看不出花纹效果，因而手上控制便使它存在着一定的偶然性，浸染结果显得较随意自然。即使是同一个手工艺人制作几块同样的扎花布，纹样也不可能是一模一样的。扎染产品的花纹、色彩如何，与制作人的制作经验、技艺高低、对纹样成型要求的理解、扎花线的松紧和染布时的浸染次数造成受色多少等有着直接的关系。蜡染在花纹的制作控制上比之扎染要容易些、人为化些。由于蜡染均是采用先平面打好形亦即用蜡涂绘的方法，染前可以直观地预知染后效果，加之有蜡的部分绝不会染上颜色，制作过程好像画画一样，因而在纹样的把握上比较容易，所以蜡染在纹样上显得比较人为化。从浸染出的纹样效果看，扎染的纹样比较柔和、朦胧、自然，在白色与蓝色花纹间能产生出较大范围的非均匀渐变层，纹样边缘较模糊，仿佛是一朵朵潜蕴在水中或雾中的花，既明丽又含蓄。传统蜡染的花纹一般比扎染清晰，也更为规整。由于蜡染皆经过精细的手工蜡绘，蜡能将染液同布料明显地分隔开，因而染出的花纹边缘较明朗，与蓝色背景能产生极为鲜明的对比。点蜡花——手绘的另一特点是可以任意发挥想象力，画出复杂多变的画面，使蜡染呈现出比扎染更精致更奇妙的花纹图案或内容丰富的画作。扎染、蜡染各有个性，它们是我国民族民间传统工艺的引人注目的品种。民间性与社会性，乡土化与艺术化，使它们经历了较长岁月，至今仍长盛不衰，将永远充满生命活力。

民族服饰中的刺绣工艺

每逢节日和喜庆日子，在少数民族地区和汉族农村，常可看到这样的景象：妇女的包头、帽子、衣服、围腰、裙、裤和鞋上绣着数量不一、风格各异的各种花纹；有些男子的衣领、衣兜、挎包、鞋子和草帽带也绣着五颜六色的图案；居家请客时，堂屋挂着绣花喜字，檐下悬挂大红绣彩；新婚男女新房的铺笼帐盖上，瑰丽耀眼的绣花传递着喜盈盈的气氛；寺庙中上香的信徒，把绣了数月的求吉求善的佛衣、红彩、桌围、幡幛等献到大殿神龛香案上……绣花，环绕着人们的衣食住行，渗透到人们生活的各个方面。因为有这种技艺，人们把观察到的自然界的种种神奇的美用彩线直接描成画幅，或者通过巧思、转换与创造，构成千变万化的纹样保存下来，长时间地欣赏。人的形象和故事，人间万事万物，用这种技艺，能比之于平面绘画观赏效果更佳地把它呈现出来，丰富生活，使人们在生活中获得更高的物质精神享受。刺绣技艺是云南各民族服饰及生活装饰艺术中不可缺少的一个重要部分。

刺绣技法与用料。

云南民族刺绣工艺源远流长。从晋宁石寨山等处青铜文物上人物形象的衣上有花纹的情况看，汉、晋时期云南可能已经有了刺绣。唐、宋时刺绣已十分发达。刺绣用途广泛，用于衣冠装饰者最多，古今情形大致相同，它们有的绣于局部，有的绣遍全身，纹样花色各民族不尽相同，有多种技艺与格调，从刺绣技巧上划分约有以下几类：

（1）挑花。挑花又叫挑绣，是用针挑起地（面料）经线和纬线，把针上的线从经、纬线下穿过去，循环往复，绣出各种图案的一种较普遍的刺绣方法，挑花图案能与地（布料）融为一体，乍看仿佛是织布时织成的，彝族的裹背、麻布围腰、背包、伞套等常用此法绣制图案。

（2）顺针平绣。顺针平绣即用比较一致的顺向针施绣，绣面平滑，无凸凹感，这类绣法用得较多，是普遍使用的针法，特点是用针成平行状排列，有直排、横排、斜排等，构造均匀整齐，不露底，不重，顺针一般用来绣花草、树叶等图，也常用于大面积的打底，大量普通花面多半用此法绣成。

（3）立体绣。立体绣又称疙瘩包筋绣，一种是在平绣画面上以重针法和摞搭法造成结瘩状花；一种是在绣线下衬心子（如裹圆的细布条，布包棉花包线、棕丝、牛筋山草、碎布等）使绣面或绣脚纹路凸出；还有一种，用金银丝扭绣或并排绣，上加珠翠，立体效果极佳，白族挂包、挂彩、桌围等常用此绣法。

（4）乱针绣或错针杂线绣。乱针绣指用针的方向不定针次无序地交错，重叠，是种不规则的，较特殊的绣法，绣鳞、毛（鸟、禽、兽）和花卉时为表现它们的质感和动感常使用。汉族、白族、彝族刺绣中多见。

（5）贴布绣。贴布绣指用布剪成一定花纹图形，贴于底料上，沿边锁口和绣纹于贴布。德昂族、彝族和拉祜族苦聪人用各色布块拼缝为衣，拼缝处加绣花纹，和贴布绣较接近。

（6）剪空内贴布绣。剪空内贴布绣与贴布绣相仿，是将绣地（即面料）剪出空心花形，内衬以布料补严沿边锁绣之。外形似补花，有凹陷感，与贴布绣的凸出感适成对照。

（7）染绣结合。白族用扎疙瘩染青花布（扎染布）做底料，苗族用蜡染花布做百褶裙，于其上加绣，使平面与凸出的纹色相映，造成双层画面效果，雅与艳相交、和谐自然。傣族孔雀舞服和象脚鼓衣，用绘（印）、绣结合的方法表现羽翎纹色，异常绚丽。纳西族东巴戴的五佛冠，在底料上绘（印）、绣结合，画出神灵形象，既牢实又庄重漂亮。

（8）连物绣。连物绣又称裹物绣，这是一种比较特别的绣法。傈僳族在海贝上钻孔，缘孔眼将海贝缝入绣纹内，缝孔的针脚规则有序，多呈花纹状。阿昌族将忍冬花秆的心子切成圆片和雕成小花形状，以彩线顺边拉网似的缝入绣面，作为衣服和挎包装饰。各地有用豪猪毛、兽骨磨制品、铜钱、金属环、彩花石磨制品玉佩螺、蛤等连绣于挎袋、冠服上的。形式多样、变化丰富，特有的立体感具有与众不同的艺术效果。

刺绣运用的材料也比较丰富，一般用作绣料的棉布、麻布织锦等称为"地"，而绣线则多用丝线、棉线，其他还有头发、马尾、什锦线等。刺绣的花纹显示方法与织锦不同。织锦的花纹是以彩色的经纬线交织而产生的平花，而刺绣则是用针引线以不同的运针方法呈现不同的色彩，在绣面上产生高花，具有一定凹凸立体感。刺绣的工具比较简单，一般常用的工具有绣绷架、绣剪、绣针等。

色彩基调。

色彩在刺绣中占有特殊地位，云南各民族对于色彩都有不同的喜好与崇尚。汉、晋时期，永昌郡（今保山）的一些民族以猩猩血染朱（红色披毡），用白色的桐华布覆盖死者，然后缝成衣服给死者穿用，说明当时这些民族崇尚朱（红）、白二色。唐、宋时期，以绛、朱、紫色为贵。元、明以后，文献对各民族所崇尚之色做了较详细的记述。例如明、清时彝族刺绣以青色作蓝天，红色作底，黄色作龙，对青、红、黄等色彩赋予了特定的含义。这种通过视觉引起的对色彩的想象是一种常见现象，它源于色彩在物象上的反映。经过光的反射、折射、吸收等作用，相互影响，相互作用，发生变化，通过人的视线产生感觉引起各种联想。这种联物想象受物质条件的制约，于是便会出现对某种色彩的喜爱与偏好，以致出现神秘追求欲。例如，古人以稀者为贵，因而常将不易绣的并且无法多得的物色列为上品。即使是黑、白、蓝、青色等举目可见的物色，由于众人皆有，大家都爱，也会不知不觉地约定俗成，变为社会尊崇之色。直到现在，各民族对色彩的审美感觉和崇尚还一直保持着古代沿传下来的若干传统的影响。例如，白族喜欢白色、青色，常以白、青为底，用对比强烈的净色（黑对白、青对白）作素绣，因为认为白是清爽、纯善、光明、高尚、庄重、吉利的象征，青色代表希望、纯朴、实在、感情真挚之意。青对白即清清白白、光明磊落。彝族、纳西族、傈僳族喜好黑色，认为它带有高贵、勤劳、旺盛、牢靠和待人真诚的含义，故而喜爱黑衣、黑底加白线素绣和黑底彩绣。从崇尚色彩和用色习惯上看，云南少数民族刺绣主要可分为三种色彩风格：

（1）鲜艳。彝族、白族、纳西族、傈僳族、哈尼族、苗族彩绣，多以白、黑为底，喜好大红配大绿，色彩对比极强烈，反差较大，另以中性色相间杂，起过渡或提亮作用，使纹样变得新鲜、醒目、生动。在色轮上，红和绿互为补色，等量并用于刺绣，可使红显得更红，绿变得更绿，经补色对比，使色彩鲜亮饱和。在强烈反差下，还可使人在视觉上产生一定错觉。除红、绿的冷暖对比外，还由于暖色显得高亮凸出，冷色较低沉凹缩而产生远近、凹凸的立体起伏感，更由于冷暖、明暗的层次对照和空间远近的对照而产生纯真质朴的美感。这都是和这些民族质朴爽快、刚健英武的性格相一致的。

（2）素雅。白族、彝族、壮族、哈尼族用单一色在深或浅色度的底料上挑绣，借助色相明暗对比作用，使纹样清晰突出。例如黑白绣，用黑色与白色相互衬托，白以黑为底，白因黑的反射而凸现，黑亦因白的衬托而分外醒目，因而产生鲜明的立体视觉。又如素绣，白族头巾、手绢、围腰在青蓝布或雅布、士林布地上绣以白色，两色纯度均强，色相一明一暗，色泽一亮一沉，相互对照，呈明快色调，异常洁爽。白族认为，青蓝表示洱海的水，白色是苍山的雪，玉洱银苍儿女的审美情趣与艺术想象，仿佛凝结在这类刺绣品中。素绣是与彩绣相对而言的，虽只是用单一色即一种淡冷色线绣，

彝族刺绣

但是仍蕴寓着"彩"味，即所谓"素中带艳"的意境，使人不仅不觉单调平淡，反而会因宁静素雅之色而受吸引。白族、苗族蓝白色花布染底加绣，也是如此。

（3）灰淡。灰淡的基调是同色绣或浅色线绣。一般用浅色线绣同色深地或反

之，暗色调者较多。哈尼族青布衣用同色挑绣花纹，经洗涤后，绣线褪色，略浅于地，纹样隐约而现。经过一次次洗涤，面料和纹样都会相应一次次褪色，色彩逐渐减弱后，出现朦胧的色纹，待衣服褪旧后，再重新染一遍，又反复上述情形。彝族、白族、壮族、纳西族等民族有在浅底上加同一深色或浅近色纹绣，经用水洗涤，深色绣脚处现出色渍印，隐约现其隐纹，好像纹变双层，出现特别效果。灰淡色调含蓄、神秘，见层次于迷蒙之中，蕴含着粗朴恬秀之美。

纹样种类和文化内涵

云南各民族刺绣纹样种类繁多，不胜枚举。纹样的题材、造型、色彩等各式各样，让人眼花缭乱，目不暇接。表面上看似千变万化，仿佛具有魔力一般，但实际分析起来，即会发现其大多数均取自客观原型，大多来自生活与自然，是依据各种原型先创造一种基础纹形而后逐渐演变、扩展、变形而来的。仔细琢磨，各基本纹形的原型依据，大致可分为三个大类：

（1）纹样中有许多是表现人居环境中的自然物，及人对它们的认识和想象，它们渗透着人对大自然的感情。这些自然物包括：

植物类：蕨类、草蔓、葛藤、树木、花卉、农作物果实等。数量以花卉最多，如蕨菜花、茨菇花、刺花、刺叶花、八角、野团菊花、素馨花、石榴花、大菊花、水仙花、玫瑰花、粉团花、莲花、茴香花、芭蕉花和其他杂花等。纯粹用花（包括蓓蕾、花朵、花瓣并以花的茎、枝、叶、叶蔓和花盆、花瓶、花篮、花窗作陪衬）或以花为主组成图案的刺绣很多，如彝族和白族的头巾、围腰、飘带、背被、背带、草帽带、口弦包、鞋帮，纳西族飘带，傈僳族衣裙，哈尼族袖头，傣族围腰，笠帽飘带等。

动物类：动物类又可分兽类、畜类、鱼类、昆虫类及禽、鸟类五种。

兽类有象、虎、鹿、猴、野兔、松鼠、麂、水獭、山猫（破脸狗）、玉面灵（白鼻山猫）、豹、豪猪等及其皮毛纹色。苗族披肩仿虎皮纹色绣制，彝族和白族裹背带、帽带、飘带有松鼠或小猴攀枝、水獭、玉面灵漫游图形，小孩戴虎头绣帽、山猫绣帽等。

畜类有马、牛、猪、羊、骡、驴、犬、猫、兔等或其毛皮纹色。彝族鞋帮绣有排列状的羊群。白族和彝族枕巾、帐围、檐彩有鸡、猪、牛、羊、马五畜合绣。线绣猫头帽、兔帽、狗头鞋、双犬兜肚在一些民族中流行。狗牙形纹、猫爪纹几乎各民族皆用。禽、鸟类主要有鸡、雉、锦、鸡、白鹇、孔雀、凤凰、喜鹊、山喜鹊、太阳鸟、火鸽、绿翠、水鸭、鱼鹰、白鹤、鹌鹑、鹦鹉、燕、鹭、鸳鸯、麻雀（谷雀）、斑鸠等或它们的羽毛形状和纹色。拉祜族、德昂族绣鸡爪花于衣、裙、挂包。哈尼族绣白鹇、双鸡于挎袋、飘带。白族、彝族刺绣花纹有四凤穿花、双凤朝阳、燕来春到及鸲鹆、画眉、鱼鹰、鹦鹉、公鸡等。居家中堂喜挂彩绣松鹤、丹凤、雄鸡、喜鹊等条幅。婚事备鸳鸯枕、夫妻互赠燕双飞手巾。在衣服上，拉祜族绣鸡脚花、斑鸠张嘴花，德昂族绣鸡爪花。

鱼类有鲤鱼、细条鱼、大头鱼、团鱼、龙眼鱼、弓鱼、泥、鳅或鱼鳞、鱼尾纹。多见于围腰、飘带、鞋帮、小孩帽及桌围彩绣。居住区靠近水域的民族如白族、傣族、壮族、彝族较喜采用此类纹样。

昆虫类有蜂、蚂蚁、蝴蝶、蜻蜓、蝉、蛾、蜗牛、多脚虫、蜈蚣、水夹虫、萤火虫、蜘蛛、蟋蟀、水板凳虫、金龟子、蟑螂、蚱蜢、毛虫、蝗虫，或虫背纹样。景颇族、傣族、彝族、哈尼族刺绣有蛇和蚯蚓纹的变形。织绣蜂与蝴蝶纹的民族较多。傣族笠帽带、白族背带、彝族兜胸围腰、傈僳族围腰、苗族绣围、阿昌族头囊顶饰等所绣蝴蝶纹，均极精美。拉祜族、佤族衣裙上的蜘蛛网纹、蜈蚣纹、毛虫脚纹等，古朴自然。

天体、自然类。天体、自然纹有天、地、日、月、星、云、虹、水、河、山、石、昼、夜等，一些民族刺绣品上的条纹、曲线纹、几何纹常有天、地、水的含义。纳西族女子背的"日月七星"羊皮，绣日、月各一，"星"五颗，"星"的中央缝有二条麂皮细带，象征光芒四射。苗族戴于脖后的方绣，有天、地和日、月映照下的"古路"和"古城"。拉祜族、佤族、哈尼族织绣有太阳、彩虹、海水、山林等花纹，拉祜族的海水纹上还显现水的反射光影。藏族热巴（艺人）背挂五色彩绣与丝（绸）带，每色代表一物，分别象征天、地、日、月、海。彝族、白族裹背用色彩明暗衬托日、月纹，明为昼，暗为夜，意为朝暮吉祥，传说裹背内的孩子因此受日月精华，可长命百岁。

（2）刺绣，像人在写字，在画画，在唱歌；像在祭祀，也像在传情。无论它表现多少内容，人，特别是本乡本土的人是它要表现和赞美的核心。因此，纹样中表现人及其生活文化者较多，如：

人体外貌：人体外貌类多表现人的衣着、外形及立、坐、卧、走各种姿势，也表现人肢体、器官的全部或局部。

活动情景类：活动情景类多表现人们的衣、食、住、行等如乘象、骑马、谈话、逐兽、放牧、饲禽、唱歌、跳舞、竞渡、劳作（采摘、收割、掇拾、挖锄、挑担、背物、栽插、手工）等。彝族、纳西族、白族有众人联袂踏歌纹挑绣，二人、四人抬轿、二人提篮穿花等挑刺纹。佤族用彩线在包头一侧缀绣数人携手跳"邦背"（歌舞）的舞姿和队形。其他如双人松鼠松枝二人叙话，四人望梅，一人或数人捉鱼、放羊、浇花、摘果等纹样在各民族地区都比较流行。

文字类：刺绣中有一些文字纹如万字纹、寿字纹、十字纹、人字纹等，穿插在图案中使用。此外文句、字符在各地刺绣中也比较常见，如福、禄、寿、喜大字中堂条幅，"喜"字布帘、枕头、被面、桌围，"情如梁上双燕子，字似池中两鸳鸯"，"水照真心风传意，好花藏在刺蓬中"等寄语言情作为异性相互赠物的手绢头巾有四季清吉、大吉大利、佛命、长命富贵、天官赐福和别的经句、咒符的童帽。少数民族文字有藏文绣咒符、经义字（有些是梵文）；彝文吉祥语；西双版纳傣文、德宏傣文的祈福经语。中华人民共和国成立以来，文字刺绣纹饰更产生许多新内容，如幸福、团结、友爱、勤俭、山高水长、互敬互爱、相亲相爱等，有的直接刺绣热爱党、热爱社会主义类语句。

建筑、工具类：白族、彝族、纳西族刺绣有拱桥、木船、畜车、房舍、照壁、田棚、龙门、亭阁及田、陌、路、沟纹等。苗族脖后方绣的"城"有城墙、道路、房子、水井纹。劳动工具和生活用品如弩、箭、锄、纺车、镰刀、梭、镖、叉子、葫芦、瓢、梭、篱笆、棚栏、渔网、兽网、酒筒、斋盒、如意、棋盘、乐器、铜炮枪等纹样流行于各民族，相互略有差异。

（3）有些纹样并不一定具有明确的意义，主要用来衬托其他纹样或专门作为美化图形用的，这类纹样叫装饰纹，多由纯几何形和自由形组合排列而成。各民族皆有其独特的纹样，如景颇族、阿昌族、傣族、德昂族、基诺族的直纹、弯纹；白族、彝族的麻子点、梭针眼、压脚线等。

刺绣纹样的构造方式一般多采用单纹，即以一个纹形为一个独立的纹样。单纹多见于帽顶、围腰、飘带头、鞋跟等，也有复合纹，即多种纹样的结合或单一纹样的连续排列，如白族、彝族、傈僳族挑绣头巾、围腰、裤边和白族、彝族的"四凤临窗"等。此外还有一种构造方式是套合纹（放射纹），以一纹为中心，向四方扩展，常见的有纳西族的日月七星装饰，傣族、阿昌族妇女元青包头绣饰等。这些纹形结构虽有一定规律，但纹形组织、纹样变化却不时带有随意性。刺绣时有些先用石灰汁或墨汁打线稿（深色布用石灰水画稿，浅色布用墨汁描样），有些却不打样，而是凭心构想，线随针引，在一个基本格式内，随意绣出。透过娴熟的技法产生出各种意外新鲜的效果，把自己的聪明才智淋漓尽致地发挥出来。

自然与生活是云南民族刺绣纹样取材的来源，在种种奇特多变的纹样图案中，深蕴着民族历史文化传统，飘荡着浓郁的山野生活气息，融贯着各民族人民的智慧、汗水和感情。在仔细读识刺绣作品内容时，人们可以感受到各民族对自己民族文化的珍爱，这是刺绣色彩斑斓、纹样变化万千和充满生命力的主要原因。人们爱刺绣，把刺绣和人的活动紧密地连接在一起，古时是这样，今时也如此。无论在坝区或山区，随时都可以看到各民族妇女利用一切闲暇时间不分地点地挑花绣朵，即便到深山密林中去放牧，也不忘带着花绷子，边吆喝畜群边坐地绣花。当然，此时此刻，绣花女眼前的一切动人美景都有可能随着针线留在面料之上，把刺绣作为表达感情与美的载体，颂扬刺绣创作者自身的美好生活和本民族的文化传统，赞颂他们赖以生存的大自然。通过纹样、色彩，巧妙地概括与变形，寄寓了淳朴的感情和美的意境，让它永远倚伴着人，使云南的天地永远是一片五彩缤纷。

织锦工艺

"织彩为纹曰锦，织素为纹曰绮。"据《后汉书·西南夷列传》记载，哀牢人"知染彩文绣…织成文章如绫锦。"可见，织锦在云南也是历史悠久，源远流长。织锦在云南少数民族中，著名的有傣锦和壮锦。其次，景颇族的织锦图案丰富。苗族、布朗族、藏族、拉祜族、阿昌族、基诺族、佤族、独龙族、傈僳族、纳西族、彝族等民族也有织锦，但使用不太普遍。织锦在各民族中，虽织艺有高有低，图案有繁有简，用途也各有异，但都具有独特技法和美的风采。

织锦，是一种特殊的艺术，它不仅在构图上有特殊的要求，而且技法也是多种多样，色彩配搭、装饰布局等也都无不在艺术的构思中，织锦构图上的奇巧处，主要表现在对几何纹样的处理上。几何纹样中大量的主体图案是棱形、方形和八角形。这些简单的单元图案，经过相同形体的位移、相似形的转换等手法，图案显得十分丰富，并且明快活泼，而又不失其朴实之美。有些纹样虽然仅仅是一些点和线，由于运用"重复"和"整齐美"这一图案设计规律所产生的特殊效果，再加上粗细线条的穿插排列，宽窄疏密的变化，使得图案更加富丽堂皇，多姿多彩，并表现出有节奏的运动感。

织锦用途广泛，做衣料、做饰件、做幕帐，当艺术摆设或悬挂的欣赏品等等，任作何用，皆让人赏心悦目，百看不厌。大概由于锦的特质产生的精神效应，古代云南曾把锦视为区别人身份、地位的衣料。唐代，云南盛行养蚕纺丝，出产各类丝锦，王者、宰相以锦制衣，凡用朱红、紫色做的衣服皆为上品，尽由官贵们享用，一般百姓不许以锦为衣。锦在人们心目中的美与贵，由此可见一斑。

云南的织锦生产，在历史上曾较普遍和著名。至近现代，主要保存和延续在部分少数民族地区。傣族、景颇族、德昂族、布依族、佤族、拉祜族、壮族、苗族的村镇，20世纪50~60年代有不少织锦能人，锦的产量也较多，除自给自足外，少数尚有富余，能拿到市场交换，如西双版纳、德宏的傣锦就是这样的。随着经济发展，工业纺织品大量涌入市场，对传统手工纺织业造成冲击和挤压，民间织锦的人数骤然减少，产量大幅下降。20世纪80年代以来，一些原来织锦较多的村寨暮夜织机声已越来越少，古老手工织锦技艺面临逐渐消失的危险。

织锦主要通过手工织机织成。傣族、壮族多用木架织机，景颇族、佤族、布朗族、德昂族、拉祜族多用腰机踞织，苗族使用的织机两种皆有。织锦普遍色彩丰富，纹样种类较多。纹样主要取自本乡本土自然与生活事象原形，或仿形，或经加工提炼使之抽象变形，构成各种生动的观赏性强的图案，地方民族特色浓郁。

傣锦。

傣锦是傣族妇女的手工艺术珍品。她们的傣锦图案，并不一味地去模仿自然，生活中一些不被人们看中的小虫、小花，经她们巧手的处理，或简化或夸张变形抽象为几何纹"入化"到自然之中，更增加装饰性比原来的自然更美。传统的傣锦图案有大象、孔雀、贝叶、大树、人物、驮马等，虽已抽象变形，但形象生动、概括简练。

傣锦花纹绚丽多姿、奇幻无比。花纹多采用动物、花卉、树草、房屋等形态，加以抽象变形成为较概括的几何纹样，再交错排列，产生变化多端的视觉效果。在一种傣锦中，常有以某种纹样占主导作母题的，便被以这种纹样来作为这种锦的称呼。依纹样的称呼傣锦常见的主要有以下几种：

（1）菱形纹和八角纹织锦。这是一种在生活中广泛使用的最常见的锦。在德宏傣族乡村中较多见。它以菱形几何纹连续或套合，衍化相组，或以八角形纹连续变化相组，错综配迭。以黑棉线为经、彩色线为纬织成，交错配以花草、几何形纹样作衬。用色多，层次清晰，鲜明富丽，庄重美观。也有用两色织成的菱形纹、八角纹素锦，简洁雅致，十分耐看。此外，菱形纹配双鸟纹锦，亦是傣锦常见的品种，这种锦纹出现的频率较高。

（2）象脚纹织锦。纹由抽象化的相连式象脚纹组成。用深蓝、红色棉线及黄色丝线织成。色调浑厚富丽，质地厚重，用作被面可数十年不坏。

（3）孔雀纹织锦。傣族视孔雀为美丽、吉祥和善良的象征，以孔雀纹织成的傣锦，花纹变化很多，色彩清新漂亮，常是结婚志喜之物。

（4）屋顶纹织锦。以傣族佛寺屋顶及其装饰物为题材而织成的屋顶剪影式纹样锦，多祭佛使用，纹样造型别致生动，花色简明美观。

（5）象驮塔房纹织锦。以象驮宝塔房、象驮供鞍纹、诵经亭纹及奉献花柱纹组成的图案，常用于佛幡装饰，也有的用来装饰床单、枕头、靠枕、手巾等物。

（6）菩提双鸟纹织锦。菩提双鸟纹含吉祥之意，代表佛祖的菩提树和双鸟在人间护佑生灵，为傣锦常见纹样之一。这类锦多为棉质，常作床单、被面、桌布和供品使用。

（7）绕线板纹织锦。绕线板即绕线工具，是傣族妇女进行纺织活动的必备之物。有的能工巧匠把绕线板刻上精美图案，使绕线板本身就成为一件既实用又美观的工艺品。而把其美丽形象化为图案织入傣锦，更体现了傣族妇女的丰富想象力和创造力。

（8）人、马、船纹织锦。织锦把人、马、船的形象巧妙变形组合，融为一体，这样既表现人行船、放马的真实生活，又包含着前往佛寺献佛行程的意境。其构图主次分明，想象独特，为傣族佛幡用锦的装饰纹样之一。

（9）驮花马和跑马、跪马、牵马纹织锦。这种锦纹样同前述人马、船纹含义类似。以马驮花表现人对佛虔诚。由于马是生活中常用牲畜，故这类锦使用广泛，会织的人较多。纹样造型在织造时常因人而异，多有变化，乡风土气和人情味很浓。

此外，傣锦还有以棋盘纹、花鸟纹、大象纹、象鼻纹、神兽纹、龟纹、螃蟹纹等为主题并混合其他纹样的织锦。各具个性，典雅艳丽。

独特的色彩和纹样构成傣锦的独特风格。它的内容和形式呈现着傣乡天地、山水、村寨、河流、森林、花草、云霓等融成的如画的自然美景和傣家富于诗意的衣、食、住、行、信仰、风俗等悠然如真的生活情景。傣锦不但是出色的傣族民间艺术品，而且是傣族生活与文化的缩影。

景颇锦。

景颇锦是云南民族织锦中独具风格的一种。无论谁到景颇山去，印象最深的可能都是用它缝制的筒裙，色彩艳丽，花纹丰富，穿在姑娘们身上，端庄华贵，显现出一种醉人的艺术效应，格外引人注目。

景颇锦的用途除了缝制筒裙外，还用于做挂包、被面、枕头、腿套和挂饰、摆设。景颇锦制作材料与傣锦相同，一样是传统的棉、丝、毛质经纬彩线。在织造技术上主要用织机（俗名腰机）织。方法是织造者席地而坐，经纱的一端绕辊，用脚掌顶住，或者将经纱头拴在地桩或房柱上，另一端系于腰际，双手用木刀引纬打纬而织，俗称踞织。这是比较古老的织造方法。现在的踞织法较以前有所改进，有用细竹签或棕制成线综装置，提升经纱，形成织口，引纬织之加上光滑的竹木挑纬刀的作用，使花经与地经分开，用杼贯以各色纬丝，不但能织90度的平纹，还能织斜形线花纹。功效虽慢，产品却结实、自然、华美，比古老踞织法效果好得多。

景颇锦色彩像朝霞灿烂，似焰火炽烈，浓艳、瑰丽、异常华贵。织锦主要有两大类型，一是黑地红纹锦，这是产量最多的一种。地为黑色，色调浓郁，花纹用朱砂偏暗红为主，适量点缀其他色，使红、黑对比强烈，再配以亮丽的黄色点缀，主次清晰，层次丰富，色调浓重。加上棉、毛质的天然织纹与绒面感所造成的折光变化的特殊效果，使其尤显庄重、浑厚、富丽，别具一格。另一种是底为蓝或黑色，质如棉布，在大片蓝或黑色底上以适当间隔织上条块状彩色图案，以黑托彩，典雅瑰丽，艺术味很浓。

景颇锦的纹样大多采用几何纹、波浪纹、连续纹等穿插组成的构图方式，较概括抽象。

景颇族的筒裙，是不同于傣锦的另一种风格的织锦。它用几何纹织满底的手法，以红色为基调，黑色为底衬，图案对比很强。景颇族妇女穿的织锦筒裙，所有花纹不但有名字而且有含义，正如谚语所称："筒裙上织着天下的事，那是祖先写下的字。"聪明的景颇妇女，在他们进行艺术劳动，对着经纬线仔细计算的同时，融进了对大自然的赞美，对生活的热爱，对未来的憧憬和对爱情的祝愿等丰富的感情。所以，有人说，景颇族的织锦可以传达感情，可以翻译成赞美自然的抒情诗。

佤锦。

到阿佤山见到佤族织锦人，差不多都有一个感觉：佤锦像一道道彩虹，既美丽，又自然，每一幅图案都像是佤族同胞居住环境和生活情趣的写照，让人过目难忘。佤锦的用途较多。首先是用于缝制妇女穿的筒裙，其中供少女穿的五彩缤纷，色调十分鲜艳；中老年妇女穿的色彩层次变化较少，色调也稍暗一些，其次用于做挎包、裹腿、被盖、床单等。佤族男女都用佤族挎包，挂在服饰色调单一的男子肩上时，由于反衬作用，显得十分富丽。沧源、西盟和孟连等地佤族妇女过去普遍善织各种花纹图案的佤锦，后来由于工业纺织品的普及，人们对锦的需求已逐渐下降，因而经常织锦的人日渐减少。

按质地分，佤锦主要有三类。第一类是麻质的，全用麻线织成（佤语叫"竿星"），花纹以直线纹为主，白底上加红、黑色线纹，色调和纹饰比较简单。这是佤锦的原始类型。第二类是棉锦（佤语叫"竿歹"），以棉纱线为主织造，色彩纹样多变，花纹除使用染好的棉质纱线外，还穿插些麻、毛质的线和仿金、银线，使图案显得鲜丽、灿烂。第三类是混合锦，有棉麻混合、棉麻丝混合、棉毛混合数种。由于采用多种材料，色彩、纹样选择余地增大，因而锦文和色调比较自由。

佤锦全用腰机坐织。织者席地而坐，把经线一端缚于房柱上（或房侧树上），另一端挂在系于腰部的宽皮带上。用若干细竹棍按规律挑起或压下经纱，挑出织孔，用梭引纬穿过织孔，拉直，然后用穿过经线的梳板将纬线打紧，如此循环往复。织的速度虽慢，但是线匀孔密，质量颇好。

花纹典雅漂亮是佤锦的一大特色。佤锦花纹中出现最多的纹样是雀眼睛花，佤语叫"艾醒"，其基本形状是在菱形框中加点，使"眼皮""眼珠"俱全，织成仿佛睁着的雀鸟眼睛。此外，还有老虎脚花（佤语叫"姜斯外"）、几何纹绕花（佤语叫"果格"）、弯形框格花（佤语叫"咯"）等。

佤族同胞世代居住在山区，与森林、泉溪、雀鸟、野兽、花果等关系密切，同大自然建立了深厚的感情，这一点在织锦上也有较清晰的反映。按民间的说法，锦的色调，底为黑色，表示牢靠，就像依靠着大山森林，是享不尽的衣食之源；花纹中的绿、红、黄、紫色等也与植物和动物密切相关。锦的纹样——雀眼睛花源于人们和花鸟的接触、观察和由此产生的对它们的好感。据佤族老人说，雀鸟比地上的野兽聪明，本领更大，因为它有一对翅膀，可以上天。雀眼睛不仅代表智慧，而且由于它看得又远又宽，还是天地日月的象征，它是吉祥的标志。有了它，上山打猎，下地干活，不管天晴下雨，都可以免掉灾情祸事，人眼若像雀眼一样亮，世上就没有什么可担心的事情了。

云南民族织锦，宛若云霞，绚丽多姿，映衬着绵延不断的山水。各族人民的聪明才智和巧技，把千里边陲大地装点得多姿多彩。

第七章　服饰与文化

要深入了解服饰文化，必须从社会文化入手。服饰不仅仅是简单的御寒防风蔽身之物，它还融汇了人们对历史的回忆、对社会的认识以及对未来的展望。它既体现了人们的世界观与审美情趣，也反映了生活准则与社会伦理观，涵盖了社会文化的各个方面。

年龄与服饰

不同的年龄，人的社会角色与家庭身份也不同。年龄与角色变异在社会中基本一致，所以，同年龄的人群，其社会角色也一样。年龄与角色差异往往也体现在服饰上，也就形成与一定年龄段相应的服饰变异，并由此形成不同的服饰类型。人的一生，在服饰上的变化可以分成婴儿、儿童、青年、中年与老年几个阶段。服饰色彩的变化，是最普通的一种变异形式。儿童服饰的色彩，多鲜艳活泼，形式简单。青年人的服装，讲究色彩明快，线条流畅，精工制作，达到服饰艺术的较高成就。进入中年以后，鲜艳的色调逐渐为素雅清淡的色调取代，老成稳重成为主要仪态。老年人的服装，则更为平淡朴实，多用暗色，也不再讲求美丽的花纹图案。衣式、饰物与搭配是不同年龄段服饰变异的主体。婴儿服饰一般都简单，也很少体现男女差别，轻软保暖与透气是制装的主要标准，饰物很少使用。童装除体现男女区别、与婴儿服装不同外，还凝聚着母亲的一片爱心，把对孩子的爱倾注在服装上，所以，童装制作精细、色调活泼、引人注目。童装还有一个特点，它包含人们对幼小生命的关爱与期望，避邪吉祥等健康美丽的图案设计特别多。花、鸟、虫、鱼、十二生肖与神佛像、长生保命、喜、寿、福、禄等花纹图案与文字就是明显的例子，这类图案多集中在鞋、帽与饰物上，使童装这几部分的文化内涵与艺术成就都特别突出。儿童时期多数从3岁左右开始，到十三四岁或十五六岁不等，时间较长，服装的积存也多，在云南的民族服饰中，无论在数量上还是在种类与文化艺术成就等方面，都有相当的分量。成年礼在过去是许多民族都举行的神圣仪式，换装是其最重要的内容，现在仍有一些民族保留成年礼，做繁复的礼仪。不举行仪式的民族，则把换装看成成人标志，到一定年龄，换上青年人的服装，就算进入成年，可以参加青年人的社会活动。成人装是成熟固定的衣式，可以保持到老。

从童装到成人装，除衣式改变外，一些标志性的衣饰变更，在许多民族中都普遍存在。在纳西族摩梭人中，换裙子是女子成年礼的重要内容。换头饰是成年换装中最常见的一种，基诺族女子成年，要改发型成独辫，并围围腰，上衣背心用成年装；基诺族男子主要用帽与太阳纹花的形状的改变来体现，成年男子会把童年的帽子换成包头，花纹也只做成圆形，不像孩童时可圆可方。泸西、弥勒等地彝族大黑彝支系，成年换帽是众多换帽礼中的一次，大黑彝女子在幼时戴船形帽，之后改冠子帽与长尾帽，16岁以后才换成成人帽，婚后还要改。蓝靛瑶女子顶板包头是成人标志，男子则是帽换帕，包头帕是成人标志。墨江等地的哈尼族碧约支系，成年女子最重要的事是制作两顶帽子，幼时只一顶，成人两顶，是因为成年少女都要参加社交活动，青年男子往往取走其帽以示爱慕，不戴为失礼，为防老人询问，只能备制一顶，以防不时之需。红河哈尼族彝族自治州一些地方的彝族、哈尼族，编辫是成年的重要标志。元江等地哈尼族糯美支系，成年后要在腰上加饰哈尼语称之为"批甲"的饰物。

成年之后，服装式样已经固定，很少有改变，而一些饰件与标志性饰物则在婚姻与生育两个关口面临新的更换。一般来说，婚装是青年服饰的最高水平，也是服饰艺术的顶峰。它仍然属青年服装。婚后坐家，或在生育之后，角色与地位已从青年的单身贵族身份转入另一种角色。女子婚后就得在衣装上做一定的标识，以区别于未婚青年，以适合于自己的身份，避免不必要的麻烦。这种变更，以头饰变更的居多，有的是全部更换，如帽子换成头帕、包头之类，也有的只作局部更改，主要是改变头饰的形状，或去掉部分部件。马关县的傣族、石屏县的傣族、金平苗族瑶族傣族自治县的瑶族支系红头瑶等，就是这类改变的例子。已婚未婚，头饰包扎法明显不同，一看便知。彝族阿乌妇女把婚前的鸡冠帽改为婚后的勒勒帽，撒尼女子在头饰上的改变也属这一范围。德宏傣族由裤改裙，也是一种标识办法。

发型发式的改变，也是一种办法，常见的有披发改髻、改辫，单辫改双辫，或多辫改髻，或改变发髻式样等许多种，这类改换在多数民族中都有。

腰带或其他一些饰物附件，也会用于标明已婚妇女的身份。哈尼族白宏支系的妇女，婚后有专门做的腰带标明身份，哈尼族称作批次。奕车新媳妇，则用一种特别的篾帽，向路人表明自己的身份。

饰品的收藏、减少，也是已婚妇女衣饰的重要变化，年轻时佩戴的饰品，婚后逐步减少，收藏起来，以备传给后人。已婚妇女身上饰物太多，难免招摇，与自己的身份不符。

老年人的服装，较中年人更为简单，注重实用，不重花哨。进入老年的标准，不全在于年龄，还要看个人的地位与身份变化。一旦儿已生子，子孙绕膝，即便年纪不是太大，着装和行为也要与身份相协调。有的地方，老年人服装的特殊标志，可以穿丝绸制的寿衣，是普遍的例子。石屏县牛街乡的彝族，已有孙子孙女的女人，过去还套一种专门的头帕，以区别于中年人，如此之类很多。

婚丧礼仪与服饰

婚装与丧服，在服饰文化中有特殊的意义。婚礼是人一生中最荣耀、最幸福的时刻，人人都要以最佳的仪态出现在自己的婚礼上。心灵手巧的姑娘，通常都要花很长的时间准备自己的嫁衣，好在婚礼上一展自己的姿色，展示自己精巧的手艺。女方家庭也很重视打扮新娘，会尽其所能装饰新娘，既示爱女之心，也要表现家庭实力。所以，过去的新娘装与新郎装都格外奢华、夸张，尤其是新娘装，简直就是一件件精妙的工艺品，代表了每个人的最高水平，饰品的搭配也最完整。

新娘头帕是一个女人一生只用一次的衣饰，最初大约来自汉族。许多民族都有新娘头帕，有的只是一方红布，也有人会绣一些花，现在已不再流行，逐步退出了历史舞台。

丧葬礼仪要求有专门的服饰，一类是让死者穿上装棺的寿衣，一类是披麻戴孝的儿孙子女的孝服。寿衣一般都要提前制备，多用丝绸布缝制，有的地方忌用缎子，因缎音同断，有断子绝孙之意。寿衣的衣样，有的同一般的老年人服装，也有的有特殊的样式，一般都不重绣花。寿衣同一般的衣服一样，衣、裤、鞋、帽要一应俱全。装棺着衣，各地风俗不同，衣的层数也不同，有的地方要求男九女七，有的地方则计几件，以多为荣，有的地方把婚装与寿衣视为一体，专门制作，一生只穿两次，一次在婚礼，一次在死后作为寿服。

孝服只在服孝期间才穿，礼毕就弃，制作也比较简单，有衣、裤、鞋、帕等，通常都用白布制成，只有忌白孝者才改用其他色。孝服只限死者的儿孙子媳等孝子穿，其余参加葬礼的人，只戴孝帕，别无装饰。

爱情信物

服装与饰品、附件中的一些特定物品，在过去的年代，常被用作传递情感的爱情信物。有的是感情达到一定程度而相互赠送的礼物，也有的是定情信物。石屏县与建水县南部交界地区的彝族，女青年会做衣服裤子等物送情郎，绿春县的哈尼族腊米人则有专送情人的特制衣服。哈尼族卡多支系，有的地方的年轻女子，在围腰的右下角边绣有一种特别的花，称作约花，是单身女郎的标志，等待有情郎取走这朵花。采花示爱，开始相恋，如长期无人过问，还会招人笑。绣的花有山茶、牡丹、杜鹃等。石屏县牛街乡的彝族，过去流行送女子亲手绣制的花鞋给情郎作信物。元江哈尼族彝族傣族自治县等地的哈尼族有送腰带定情的，彝族也用腰带作为爱情信物，腰带绣制精美。拉祜族苦聪人，有的地方用黄、绿、蓝、白、红五色彩带当作送男子的信物。男子通常回送丝线、手镯或其他的工艺品。德昂族男子送腰箍给女子作信物。基诺族男子则送自己编的饭盒、槟榔盒、针线盒之类与女子，女方回赠腰带、草帽带、绑腿、刀绳等物为礼。

绣制精美的褂子是一些地区彝族、哈尼族的爱情信物。傈僳族有专送男子的两种小包，一装火药，一装钱物。送了之后，女方要考验男子看他能否用袋子中的钱做生意获利，用火药等物获取猎物，这一习俗主要流行于德宏傣族景颇族自治州。

节日与服饰

节日期间是各族群众展示手艺的好机会，人们都要穿上自己最漂亮的衣服，挂满各种美丽的饰品，节日往往成为一场场一次次的赛装会。彝族的火把节、白族的三月街、景颇族木脑纵歌、苗族的花山节、傣族的泼水节、壮族的三月三、傈僳族的刀杆节等，无一不是争奇斗艳的服装盛会。节日期间还对服装有一些特殊的要求。西双版纳傣族过泼水节，只扎红、黄、白三种布条，不包头帕。富宁县的彝族过跳宫节，宫头服与宫主服都是特制专用，平时不穿。弥勒市的彝族祭火，只用草叶、木叶蔽体，景颇族在埋魂仪式中也不着布衣，彩涂之外，还加树叶等物蔽身。这种活动，既有追溯古仪的意思，也有驱恶除邪之意，并且只限男子参加，无伤大雅。

许多民族都将服饰分成便装与盛装两种，便装是平时劳动休闲时穿的，盛装则是专为节庆等重要场合准备的，有的地方，便装与盛装的区别仅在有无饰物上。所以，要见识传统地道的民族服装，节庆是难得的好机会，不仅穿得整齐，人也多，各种各样的衣服都会在节日期间展示出来。

说到服装展示，大姚县昙华山一带的彝族，还真有名副其实的服装展示节日。各村各寨的姑娘都可以在节期一展自己的聪明才智、容颜装扮，既可博得众人喝彩，还可争得男青年的爱慕追求。这个节日现在已很出名，称为赛装节。戏服与舞蹈服装，是一种特殊的服饰，往往也能在节庆活动中见到。藏族、傣族、壮族、彝族、白族等许多民族都有戏班，有成组的独特戏服。彝族、白族的舞龙、舞狮，傣族的孔雀舞，汉族的关索戏也在节期举行，都有独特的衣装与面具。这些在云南的民族服饰中占有一定的地位。

服饰与传说、故事

在云南的许多民族中，服饰是文化载体，一件简单的绣片，一个并不复杂的图案，都可能潜藏着一个动人的故事，有一个不同寻常的来历。服饰藏载的故事与传说有两种来源，一是依形拟意，根据图案的形象，附会一个故事作为解释；另一种以图记事，把历史传说与事件用一定的图案表现出来。以内容划分，有爱情纪念故事、吉祥传说、救助纪念、两性关系、先人事迹纪念、神赐、伦理故事、模仿与象形、图记史事、禁忌故事几种。

景颇族卡苦包头的来历、苦聪人的彩带来历、撒尼人的头套、壮族围腰的来历、基诺族太阳花的来历等，都把服饰中的某一部件的来历归结在一个美好的爱情故事之上，将之看作吉祥美好的象征。

布朗族的三尾螺饰品传说，壮族孔雀帽传说，彝族、哈尼族的鸡冠帽传说等，则属吉祥物类型。传说曾有一个神奇的三尾螺，使一位布朗族姑娘变成了绝色美女，脸上的肤色还能以一天不同的光线与天气发生三次变化，因此被称作婻三飘，意即一日三变的美女，后人因为有这样的传说，将三尾螺视作能使人变美的神物，制成饰品装饰自己。壮族小孩的孔雀帽，则是因为孔雀啄蛇救了一名叫狄媚的小孩，父母认为孔雀是吉祥之物，能免除灾祸，就绣成帽让她戴，后人仿效，就成了风尚。彝族中的许多支系都戴鸡冠帽，都赋予它吉祥美好的含义。居住在弥勒、泸西等地的阿乌支系，传说他们之所以戴鸡冠帽，是因为鸡曾啄死害人的蜈蚣，救了先民，认为是吉物，就将其装饰在头顶上。传说鸡与蜈蚣交朋友，蜈蚣借鸡的角去做客，一借不还，结成仇怨，鸡见蜈蚣就追啄，成其天敌。后阿乌村寨受蜈蚣精骚扰，苦不堪言，偶得人告知，鸡能克蜈蚣，就大量养鸡，鸡啄死蜈蚣精，人得安居，故用以为装饰。红河县乐育乡一带的哈尼族，则传说鸡是魔鬼的克星，所以将鸡的形象制成鸡冠帽，戴在头上，视为祥物。

有一些饰件，则被归结为救助人的纪念品，因其救助过先人，视为吉物，世代相传。傣族的船形鞋，被追溯为船渡人于水难，改形成鞋。佤族大手箍，则传说是受耳筒避过人熊的启发而制成的。红河等地哈尼族着木屐，也传说先民曾靠木屐避过敌人，后人沿以为俗。而有人把哈尼族衣色偏重黑色，也说成是先民曾经因着黑衣而躲过追敌，后人遵以为俗。巍山、南涧、祥云、弥渡等地的一些彝族，后腰有一块圆形的布或毡作饰物，上有两个类似眼睛的装饰点，据说是表示蜘蛛。传说先民被敌追，躲入山洞，才一入洞，就有蜘蛛在洞口结网，追敌不能进洞搜索因而得救，为感谢蜘蛛救命之恩，就将它绣在布上作为饰物，流传到现在。

两性之间争地位高下，具有悠久的历史，服饰中也渗入这类的内容。在白族的传说中，之所以有围腰，是怕女人太过精明，使两性之间的分工与平衡被打破，制围腰蒙其心，以平抑女人的心智，使男子居于主导地位。普米族则传说，原先女子很聪明，男子却很笨，神仙把男人的愚笨归结为女人太精明，就让女人戴首饰，玩物分心，心智被封，男人才聪明了起来。相对而言，元阳县哈尼族的传说，则赞扬了女人的才智，把每个人胸前的别针说成了与男子斗智的战利品，男人赞赏女人的才智，送别针为纪念。

有一些服饰部件，则被说成某一偶然事件的遗存。巍山一带的传说，包头是从皮罗阁战伤以布包头开始的。苦聪人则传说，洪水泛滥时，人们在葫芦中避水，洪水过后，天神用刀劈开葫芦让人们出来，结果伤着苦聪先祖的头，只得以布包扎，就成了包头的开始。傣族着白包头，则被说成是源自祖先为保护妇女儿童而牺牲的王子戴孝的孝帕的沿用。凉山彝族男子头上有角状的装饰，称为天菩萨。传说来自一个叫阿里比日的人，他因吃了龙肉，身上长出了龙角龙鳞，只能用布包裹，后人为纪念他而作同样的装扮。怒族则传说，古代有皇帝因头上长角，包布为遮盖，后人因其为皇帝，认定包头可能是力量的来源，以为吉祥，相沿成俗。

有一些服饰构件的来历则被归结为天予神赐。白族的凤凰帽被说成是凤凰给凡人的礼物。而纳西族妇女七星披肩上的日月七星，是三多天神为褒奖勇斗旱魔的女英雄英古而将日月星辰摘来镶在她的顶阳衫上，才有了披星戴月的纳西族服装，后世的纳西族妇女据说就是继承了英古顶阳衫的旧制。

前短后长式的衣服，是很多民族都有的流行服饰，但在普米族传说中却成了一桩伦理公案的见证。传说有姐弟二人，已多年未见，各在一方，不通音信，不知所在。某日弟之猎物逃脱，追至一地，被一家人遇见抬回剥食，弟追至，与男主人发生争执，其后方知是姐夫。弟耻姐家为人，虽然其姐再三劝阻挽留，仍割袂而去，留下了这种前短后长的奇特服饰，表现了一个刚强正直的民族的伦理观念，极具教育意义。

有一些服饰图案，则是一种模仿，或被因形归类，说成是来自某种相似的物品。美丽的景颇族筒裙，据说是学美丽的巴板鸟的花纹而织出来的。撒尼人头套上的彩带，则说是学七色彩虹。弥勒、泸西等地的黑彝服饰传说中，帽子是罗锅改制的，裙子是伞改制的，而衣是袍子改制的，所以帽似罗锅而裙如去骨的伞。

图形记史以苗族服饰最有特色，苗族的披肩或衣的肩背部位绣有方块、箭头、叉等多种不同的图案，据解释是描绘苗族祖居地的图案，分别表示江河、城池、田畴、官印以及一些涉及苗族迁徙的事件。昭通、昆明、楚雄等地的苗族服装，至今仍普遍保留这类图案。富宁县的大板瑶，围腰上绣有12朵梅花，据说是表示瑶族的12姓。西畴、麻栗坡等地的蓝靛瑶，头饰中有一条叫罗白的饰带，其所绣图案据说是描绘瑶族漂洋过海等迁徙历史。

赤足在云南历史上是普遍现象，近代才逐渐盛行着鞋。在元江哈尼族彝族傣族自治县那诺一带的哈尼族中，不着鞋却被说成因有穿鞋禁忌而形成的古规。在传说中，汉官的女儿与哈尼王结婚后，把地图情报藏在鞋中送给汉官，哈尼人被打败，受尽奴役，哈尼族就从此发誓不穿鞋，穿鞋也就成了禁忌。

服饰与文字

文字也是服饰的重要装饰品，吉祥美好的字句常被绣在衣服帽子上，或刻在饰物上。在汉族和一些少数民族中，"禄、福、喜、寿"等字常见于衣物与饰品上，"长生保命"等字则刻在专给小儿戴的胸牌上，起到护身符的作用。

壮族土僚支系的服饰，也绣有文字，多成句，绣在头帕和衣服的胸背部位。有汉语句子，如团结起来，满怀激情送歌献舞，鲜花求闹春，提高思想之类。也有的是汉字记壮音，还有一些是古字。

瑶族则多在罗白、伞套与童帽上绣文字，如在一个伞套上则绣有金、水、上船等字。藏族则把藏文刻在手镯等饰品上。

服饰中的吉祥图案

云南民族服饰中的图案很多，有吉祥图案、动植物图案、自然物图案与人工制品形象图案，还有几何图案。

云南民族服饰中的吉祥图案包括动物图案、人物图案、植物图案与其他一些种类繁杂的图案。吉祥图案的来源是民间信仰。云南各民族中流行的吉祥图案，有本土的传统图案，也有来自外地的信仰逐渐形成的图案。外地来的图案多取自明清以来的流行图案。吉祥物的来源比较复杂，有的来自历史传说，有的来自比喻，有的来自让人心悦的美好形象，也有的是源于同音变意。

中国人普遍认同的吉祥事例有平安富贵，多子多福，婚姻中的合和、白头偕老，为官的清廉与升迁等。这类题材是云南民族服饰中较常见的，汉族之外，彝族、白族、壮族等很多民族也有这类图案。

石榴、瓜、葫芦、老鼠、葡萄等图案象征多子，桃与灵芝示长寿，佛手与蝙蝠被借喻为福字，牡丹象征富贵，钱形示多财，莲示清廉，藕喻指偶，百合示百年好合等，都见绣于围腰、背被与衣、帽之上。

松柏长青，又在寒冬不改颜色，被奉为至诚君子。竹、兰、梅的气度风范也受人称赞，菊以不附时，被视为花之隐逸君子。松、竹、梅称岁寒三友，梅、兰、竹、菊称四君子……，是传统的观点，也影响到云南民族服饰，为常见题材。

龙、凤、麒麟为灵物，中国常见，云南服饰也常见。组图如蝶恋花、鹊登枝、鱼戏莲等也是常见图案。

十二生肖图案与造型，常见于饰品与鞋、帽，尤以童鞋为多，主要流行于汉族、彝族、白族等民族中。人物图案则以白族服饰为多见，挎包、背被都常花杂人物的图案。

大理一带的彝族、白族，会将五谷种子装入香包中，作为吉祥物，让小孩佩戴。

云南的名花如山茶与马缨花一类，也被视为吉祥物，绣在服饰上。吉祥动物除龙、凤、麒麟之外，常见的还有狮、虎、孔雀、白鹇、锦鸡、鹦鹉、鹤、鹭、鹰、雁、鸡、鸭、鹅、喜鹊、大象、鲤鱼、蛙、鸳鸯、蟹、蝶等。

云南民族服饰中的很多图案采用动、植物作底，是考虑其视觉效果。选用的植物很多，有的是刻意模仿，也有的属相似指认，认为图案像某物，就视为该物，常见的有树形、鸡枞、黄泡、萍、蕨叶、蕨芽、笋台、芝麻壳、辣椒、柳叶等。常见的动物图案有蝉、蚂蚁、野鸭、鸡冠、马鹿、蜻蜓、蛇、鸽尾、蜂巢、龟、蜈蚣等。有服装图案，也有饰物造型。

云南各民族用于服饰的自然物图案很多，日、月、星辰、雪花、云纹、波浪、火焰、水、山、石等都有。见于服饰的人工制品，多制成小模型为饰品，或做成图像。常见的有灯笼、陀螺、箭头、刀、矛、剑、铃、钟、瓶、花篮、针筒、挖耳、梭子、壶、屋宇等，数不胜数。

　　云南民族服饰中常见的几何纹有方形、三角形、梯形、菱形、十字形、圆纹、线圈、井字、T形纹、回纹、几形纹、X形纹、三角形拼图、菱形拼图、线条拼花、8字形、山形等图案。图案多，拼配复杂，拼出的图案多姿多彩，极为耐看。

第八章　云南民族服饰的保护与传承

服饰是一种"历史符号"，服饰的世代传承也就意味着文化的世代传递。民族服饰本身就是源远流长的一种民族文化，它凝聚着民族的审美意识、装饰艺术和族群特征，汇聚了地域的、历史的、民族的气息。当衣服与饰物相结合时，便在其单纯的防寒保暖的基本功能上赋予了复杂的民族群体意识和社会文化功能，成为社会关系的构成部分，进而表现出艺术美和民族精神的创造力。从经济学的角度讲民族服饰文化是一份宝贵的资源，对于区域旅游产业、文化产业和区域社会经济的发展都具有重要的价值。

随着现代社会文明的日趋普及，通过服饰体现和保留历史的作用已不再是那么的重要，一些少数民族服饰也开始出现了简洁化和实用化的趋向，其中的审美意向更趋明显，这必然导致服饰的民族性丧失。20世纪80年代，我们到云南少数民族地区旅游，看到男女老少穿着民族服装在田间劳动赶集，参加民族节日，载歌载舞，体现出浓郁的民族气息和地方特点。20世纪90年代后还看到中老年妇女儿童穿着民族服装，到2000年以后，只有老年妇女穿民族服装。在提倡保护环境、回归自然，保护民族传统文化遗产的今天，少数民族的生活习惯和服饰逐渐改变。这个问题若不能引起社会的足够重视，再过几代，少数民族服饰文化将不复存在，这将是人类文化史上的一大遗憾。

由于云南省复杂多样的地理和立体气候因素，人们迁徙定居，各个民族的文化长时间相互影响，逐渐形成本民族的独特的生活方式和文化内涵。民族服饰也形成了一套有着本民族深厚文化底蕴的传统风格。过去多数少数民族都居住在偏远地区，过着自给自足的小农经济生活，很少受外来文化的影响，生活节奏相对较慢，有的地区经济还比较落后。服饰形成自身的民族特色后，一般变化相对不大。改革开放后，社会生活发生了很大的变化。交通便捷，数字信息化时代的到来，缩短了地球的空间距离，出现了不同文化的大碰撞。这些情况影响了云南少数民族的生活方式和审美心理。

城里人穿的服装款式多样化、个性化，形式简洁大方、穿戴方便，体现着时代的审美意识。多数服装是工业化大生产的产物，因此经济实惠，人们不做服装，只选购服装。而少数民族的传统服装大多数是自己做，也有一些是由当地裁缝做。制作工艺是一代代传承下来。苗族服装满身都是花纹图案刺绣，一套服装做下来要花一年的时间，至少半年时间。有的民族服饰还要用无机原料装饰，如玉石、翡翠、宝石、玛瑙、绿松石、金、银、锡、铜。还有人工制品，如陶瓷、珐琅、玻璃等等。一套服装做下来成本也很高，从原材料的采集和处理、纺线、织布、裁布、缝衣、刺绣都是在劳动之余进行，很少有妇女能专门在家做女红而不干农活和家务事。这些都是辛苦费时的劳动。一套服装制成后穿戴也较繁杂。如打包头、系腰带、扣纽扣、打绑腿、戴装饰物、编头发等。

现在少数民族中的中青年人接受的是现代文化的教育，和外部世界的接触也较多，受到城里人的生活方式的影响，这一切都潜移默化地影响着他们的生活和审美观念。和汉族生活和工作在一起的少数民族逐渐改变了自己民族的习俗，而忽视了自身独特的民族风俗文化。如果我们不采取活性的保护机制，最后再来恢复民族服饰文化就相当困难了。

这需国家和地方政府的文化管理部门建立健全有效的活性保护机制，同时也需要民间的积极配合和参与保护。

（1）树立保护本民族服饰文化的自尊心和自信心。一方水土养一方人，本民族服饰文化的形成有自身的历史渊源和合理性，也是现代多民族文化的重要组成部分和本民族生命力的象征。那么，从民俗学的角度看，穿本民族服装，才能树立对本民族的认同感和自尊心，才能知道保护本民族文化传统的同时吸取外来文化的长处，自我发展、完善、实现自身独立存在的价值。

（2）个人和博物馆收藏民族服饰是保护民族服饰文化的一种方式。少数民族自治省、县都建立有本民族文化的博物馆。地方博物馆收藏民族文物要进行认真的整理分类，民族服饰的收藏要有鲜明的民族特色，具有原汁原味的效果，服装展览和展示设计应该取得协调统一的风格，对民族服装的文物价值、实用价值、艺术价值进行学术研究并介绍给观众。但有的地方博物馆对民族服饰的展示效果不是很理想。要实现活性的保护，需要政府的引导和支持，使少数民族在日常的生活中继续穿民族服装，并对自己的民族服饰有喜爱之心，意识到这也是现代文化独特的组成部分。

（3）由民间和政府定期组织少数民族服饰文化的展览。不仅在本地区，还要在国内，甚至到国外进行展览，让单位和有识之士进行收购。国际上一些收藏家对具有原创性、艺术效果较强的民族服饰很感兴趣，出价也不低。国内一些人士也喜欢收藏民族服饰作为室内主墙面的装饰品或收藏品。组织专家对有收藏价值和市场前景的民族服饰进行论证，确定其艺术价值和经济价值，使之有广阔的国内外市场和专业的民族服饰制作者，使民族服饰的制作工艺不断延续下去。

（4）举行展示自制民族服装的赛会，评出奖项，以资鼓励，弘扬传统服饰艺术的精神。由民间和政府组织庆祝少数民族的传统节日，要求穿本民族的服装，展示服装特色。

（5）创办少数民族服饰艺术杂志，研究其历史渊源、艺术效果、文物价值。介绍各种各样的服饰文化活动，进行民族服饰文化的学术交流。为保护活性的民族服饰艺术进行政策指导。引导民族服饰的商业流通。研讨民族服饰可持续性发展的有效方法。增强人们保护少数民族服饰文化的自觉性，丰富人们的服饰文化生活。

（6）发展地区旅游经济是保护和发展少数民族服饰文化的重要手段。随着现代经济的发展，城镇居民都喜欢利用节假日到农村"农家乐"，陶冶身心。有的人还喜欢收集少数民族的服装、刺绣、饰件、挂包、手工艺品作为纪念。

少数民族地区的风景名胜、独特的生活方式、粗犷的民族歌舞、富有特色的建筑和服饰文化、当地的土特产，都是发展旅游经济的重要资源。这些都需要当地少数民族进行挖掘、整理、发展，并加上优质的服务来吸引游客。平时穿少数民族服装能增加地方风情，还可以将身上的刺绣图案和挂件进行整理和装裱，作为旅游品出售。

云南以自己神奇的山势地貌、多样化的立体气候、神奇的人文景观、丰富多彩的少数民族形象吸引着世人，这些是我省发展旅游事业的先天优势，使云南成为旅游大省。但在现代快节奏的生活和经济发展浪潮的冲击下，我省的少数民族服饰文化已经流失。政府和人民应当立即着手建立活性的保护少数民族服饰文化的机制，制定民族服饰文化可持续性发展的战略。少数民族服饰自身需要发展，不断完善，使之充实到时代文化的长河之中，使之丰富多彩、生动活泼，与多元文化共存，也才能显出本民族顽强的生命力。

主要参考文献

［1］ 〔唐〕樊绰，向达校注，蛮书校注，中华书局2018年版。

［2］ 杨德鋆：美与智慧的融集，云南人民出版社1993年版。

［3］ 玉腊：百彩千辉——云南民族服饰，云南教育出版社2000年版。

［4］ 程志方、李安泰：云南民族服饰，云南民族出版社2002年版。

［6］ 李昆声：云南艺术史，云南教育出版社1995年版。

［7］ 李昆声、周文林：云南少数民族服饰，云南美术出版社2002年版。

［8］ 中共云南省委宣传部编：传扬生活妙韵的巧技——云南民族工艺，云南教育出版社2000年版。

［9］ 吴华、玉香：试论云南少数民族传统服饰文化的保护、利用和开发，艺术研究，2007年第3期，第14页。

［10］ 李晓弟、刘继平、朱海昆、王崇兰：论云南少数民族服饰文化的保护与发展，昆明理工大学学报（社会科学版），2005年第5卷第4期。

后　记

　　为了使更多的国内外人士认识云南各民族服饰，了解更多的云南服饰文化，我们早有向国内外介绍云南民族服饰的愿望，如今，在云南人民出版社的通力合作下，这一愿望终于得以实现。

　　对我来说，编译这样一本书是一个难得的机会，也是一件极不容易的事，繁忙的工作和知识基础不足使我不时遇到难题。然而我十分乐意去面对一切，愿意去体验一下挑战自我的乐趣。在写作过程中，我学到了不少新知识，也积累了许多经验，这对我今后的发展是十分有益的。我非常感谢云南民族大学外国语学院的领导能给我这样一个机会。此书的撰写过程是一个不断学习思考和超越自己的过程，写作过程中一直得到许多关心我的老师和朋友的支持。在此，我要特别感谢李强教授的指导、帮助和各位编委、领导的关心。没有各位的支持与关怀，这本书是难以完成的。

　　云南悠久的历史和博大精深的民族文化，为民族服饰提供了广阔的发展空间，各民族创造出的精美绝伦的各类服饰及服饰文化令人惊叹，然而，受篇幅所限，这里只能择其有代表性的进行介绍。因此，错误和不足之处也在所难免，还望广大读者批评指正。

曹霞

2021年10月

Chapter One   Long History and Various Costumes

Yunnan traditional clothing culture is characterized by its numerous clothing types, various styles, colorful colors and its unique characteristics. Cultural connotation has formed a self-contained cultural type. In the long process of development, Yunnan minority national costumes incorporate some special cultural connotations. In the seemingly simple clothes, some freshness is integrated dynamic cultural content. Therefore, clothing is no longer simple clothing, but a kind of carrier containing rich cultural content. So we call it "clothing culture".

Yunnan national costume culture can be described as "many", "abundant", "colorful" and "skillful". "Many" refers that there are many nationalities. Besides, there are many types of clothes. "Abundant" is a rich accumulation of cultural connotation. "Colorful" means rich and colorful Posture. "Skillful" means that the composition is rigorous, dense and exquisite.

Yunnan is a province with the largest number of ethnic minorities in China, with 25 ethnic minorities living in the world. In Yunnan alone, there are 15 unique ethnic minorities. These ethnic groups also have many branches. Even if they are the same ethnic group, their distribution is also relatively extensive. Almost all regions, prefectures, counties (cities) have ethnic minorities. Almost nowhere in the province, a single ethnic group is independently distributed, generally living adjacent to or interlaced with one or more ethnic groups. And the distribution of all ethnic groups in the same area is mostly three-dimensional. Taking Menghai County of Xishuangbanna as an example, Dai people live in dam areas and river valleys. Hani people, Bulang people and Lahu people live in semi mountainous areas and mountainous areas. The longitude and latitude position and the climatic zone in Yunnan are also diverse. Dai, Achang, De'ang and other ethnic groups live in the tropics, while Tibetans, Naxi , Pumi and other ethnic groups live in alpine mountains.

In addition, the social, historical, economic and cultural development of various ethnic groups is also uneven. By the early 1950s, the ethnic minorities in Yunnan may be in feudal and semi-feudal society, slave society, and some ethnic groups are even in the post-primitive society. These differences and differences have created the diversity of ethnic minority costume culture in Yunnan, showing a variety of complex and colorful pattern. In clothing culture, clothing type and style are two important cultural elements. Yunnan ethnic costume is famous all over the country for its diverse clothing styles and numerous styles. From one-piece garments related to the origin of clothing to complete suits, to the gorgeous and solemn chieftain costume and its accessories, it constitutes an extensive and profound Yunnan clothing world with a rich cultural connotation.

The achievements in technology are also worthy of praise and admiration. Yunnan Customary technique in clothing culture is that a variety of processes are integrated into one . Ethnic minorities in Yunnan like to use embroidery, tie dyeing, batik and other technologies to decorate clothes. Among them, embroidery technology is the most prominent mainstream workmanship of Yunnan ethnic costumes.

In terms of clothing aesthetics, we can also find its uniqueness. In a simple or beautiful dress pattern, most of them are endowed with rich cultural connotation, which is the most vivid and charming in Yunnan ethnic minority clothing culture.

The so-called "one outfit, multiple information" is the most vivid and appropriate description of the rich cultural connotation of Yunnan ethnic minority costumes.

Yunnan's long history, numerous ethnic groups, complex topography and changeable three-dimensional climate determine that the costumes and decorations of the ethnic minorities in the south are bound to be colorful and magnificent. There are 26 ethnic groups in Yunnan. Various ethnic groups are divided into many branches due to cultural and economic differences. Due to the different ethnic groups, different clothing and different branches, people have different clothing. Even if the same branch live in different places, their clothing will also be different. Therefore, Yunnan ethnic clothes and decorations are complicated and numerous.

As far back as 1.7 million years ago, the early Paleolithic Homo erectus, the Yuanmou people, lived in Yunnan.Early Homo sapiens, Zhaotong people in the middle of the Paleolithic Age, people from Lijiang, Xichou, Kunming and Mengzi in the late Paleolithic live in Yunnan. The mode of production of Paleolithic humans was a predatory economy: hunting, gathering, fishing. Human habitations are mostly natural caves. In order to keep out the cold, the "clothes" used by humans are trees leaf, bark, bird feather, animal skin, animal hair, etc.

In the Neolithic Age, human beings entered the era of production economy. Humans have learned to grow grains, domesticate livestock, pottery and textiles. Due to the invention of textile technology, human beings began to weave linen clothing to keep out the cold. The pottery spinning wheel found in the ancient ruins of Yunnan Neolithic is a tool for human beings to weave and make clothes. The original inhabitants of Yunnan in the Neolithic Age pay attention to decorating their body. People use bone, cornerstone and other materials to make ornaments to decorate various parts of the body. In the Cangyuan rock paintings, the heads, necks and ears of many figures are decorated with animal horns, feathers and branches. In addition, the inhabitants in the Neolithic also loved a permanent decoration - tattoos. Cangyuan rock paintings have tattooed portraits.

In the Bronze Age, the diversity and individuation of the costumes of the ancient ethnic groups in Yunnan are reflected in the bronze figures. Through different costumes and hairstyles, we can identify which nationality they belonged to in ancient times. In Records of the Historian ·the Biography of Nanyi, Sima Qian recorded the clothes of Dian people. Dian people, both men and women, wear collarless pair Clothes, knee length, no pants or skirts on the lower body.

In the Tang and Song Dynasties, because the two local governments of Nanzhao and Dali ruled Yunnan and the rank is strict, the clothing is also very particular. Manshu written by Fan Chuo of the Tang Dynasty records a lot of information in this regard. According to records, the clothes worn by the king and Prime Minister of Nanzhao at that time were all silk with colorful patterns. Ladies wore the skirt made of brocade. The bun was decorated with pearls, gold shells and amber.

In the Yuan, Ming and Qing Dynasties, the costumes of various ethnic minorities in Yunnan have shown the characteristics of blooming flowers . Especially in the Qing

Dynasty, the modern ethnic costumes in Yunnan laid the foundation of modern ethnic costumes in Yunnan.

The profound historical and cultural accumulation and the pattern of multi-ethnic culture have become the foundation of today's rich and splendid ethnic costumes in Yunnan.

At the same time, Yunnan is the transportation hub between Southeast Asia, East Asia and South Asia. A long time ago, there was a trade route through Sichuan, Yunnan to India. Trade exchanges were very frequent. Manshu and The Travels of Marco Polo are documented in detail. Yunnanese themselves, in foreign trade and inland commercial trade, also take the initiative, positive attitude.

Many ethnic groups in Yunnan have also experienced a long-term migration process and gathered here in Yunnan from all sides. Different life experiences and migration processes have brought many different cultures into Yunnan and greatly enriched the ethnic culture of Yunnan. After entering Yunnan, Han culture, which was originally a foreign culture, not only took root in the Han nationality but also affected minority areas and formed its own original style.

Yunnan's complex geographical environment, diverse climate and rich products have inspired the rich creative ability. Yunnan has a wide range of mountains and rivers and a diverse climate, ranging from low-altitude tropical rain forests to snow capped mountains above 6000 meters above sea level.

People's life changes adaptively due to different factors such as geography and climate, which is reflected in clothing.

The fur clothing of the Tibetans in Diqing adapted to the cold geographical conditions at high altitude and the lightness of the Dai people's clothing in Xishuangbanna are the best example. Generally speaking, the nationalities in cold and cool areas have long, wide and heavy clothing, while those in warm regions can be less concerned about keeping warm and prefer to have light and thin clothing, and the middle zone is also adjusted according to the conditions, forming a lot of unique clothing style.

Yunnan's rich products enable people of all ethnic groups to show their creativity and make costumes with different styles. Palm, mountain grass, bark and leaves are used as clothing materials, enriching the source of materials. The huge beak shell, the tail feathers of the golden pheasant and the beautiful peacock, the tiger claws and teeth are used as ornaments, which add strange decorative effects and customs. Yunnan's abundant jewelry also depends on the rich gold, silver, copper, tin and jade resources in Yunnan.

It is precisely because of its many advantages that Yunnan's ethnic costumes are not only diverse, but also have profound cultural deposits. People of all ethnic groups often express their feelings for beautiful things, the roads they have traveled, the things they have experienced, and the social morality they follow in the clothing culture. The simple aesthetic ideas and profound life philosophy of the people of all ethnic groups can be seen everywhere in the pieces of embroidery and clothes.

Chapter Two   National costumes and distribution

There are many ethnic groups in Yunnan, and their costumes are varied. Due to the special geographical environment and historical reasons, the 25 ethnic minorities living here each have their own unique costumes. Some ethnic groups have different costumes due to different branches. Even if they belong to the same branch, they often have different costumes due to differences in living areas. More importantly, the costumes of various ethnic groups have their own special cultural connotations in the long history of historical development. The costumes of various ethnic groups are pictures of unique style and interest, and they are also a treasure trove of ethnic folk arts and crafts. Minority costumes in Yunnan are the best portrayal of ethnic life , production methods, customs and aesthetic tastes, as well as the best embodiment of ethnic culture. Minority costumes in Yunnan are colorful and amorous. Below is a selection of some distinctive ethnic costumes to introduce.

Costume of the Vas

There are more than 400,000 Wa people, mainly living in Cangyuan Wa Autonomous County of Lincang City and Ximeng Wa Autonomous County of Pu'er City. Women's skirts are woven with five color threads. The main color is red, and other colors are intermingled. The belt and silver headband made of dominoes and five colored beads are the basic symbols of its clothing.

The costumes of Va nationality have prominent characteristics, such as headwear, body ornaments and waist hoops, which have typical national styles. The overall impression of the Va costumes is that the structure and pattern are relatively simple, and they create their own distinctive characteristics with simple lines and composition. The color of Va costumes is mainly black. Only Gengma Dai Autonomous County has a small number of Wa costumes with yellow color. Cangyuan Autonomous County and Ximeng Autonomous

County are the main popular areas of Va costumes. Some Va men wear earrings, and most of them used to wear knives. The men's trousers of Shuangjiang Lahu, Va, Bulang and Dai autonomous counties are characterized by their crotch hanging to the ankle.

Costume of the Bulangs

There are more than 119,000 Bulang people (2010), mainly living in Bulangshan Township and Daluo Town, Menghai County, Xishuangbanna Dai Autonomous Prefecture. It is a cross-border ethnic group. Its clothing is deeply influenced by the Dai people, but it also has its own style. Among them, the woven tube skirt is the most representative, and the silver ear column has unique national characteristics.

There is a big difference in the clothing of Bulang women. The young Bulang women living in Xishuangbanna and Lancangjiang wear a short shirt with round neck and narrow sleeves and a two-layer skirt. The girls have long hair, while the women wear their hair in a bun and insert a hairpin. The top of the pin is inlaid with three diamond shaped transparent glass beads, and a silver chain is tied under it. Silver pieces, silver bells and other ornaments are hung. Women's clothing in Shidian area is unique. They wear a high collar, long sleeves, large face slanted jacket, the cuffs inlaid with red and green calico horizontal strips, the high collar embroidered with exquisite patterns. They wear a silver bubble necklace. The outer cover is a calico short jacket with 15-20 pairs of silver coin buttons set on the hem. The waist of the blue cloth with rolled white edges is long to the knee. Bulang women wear blue cloth pants, tied with blue cloth leggings.The bun is wrapped with two pieces of green cloth of over 3 meters. The cloth wrapped around the head is folded into a triangle shape and fastened with a colored glass bead and a white ball. Women wear embroidered shoes.

The number of Bulang people is small. Most of them live together with Dai people, and are greatly affected by it. Due to the scattered residence, there are also differences in different places.

(1) Menghai style. It is mainly distributed in Menghai County, Jinghong City, Lancang Lahu Autonomous County and other places. There are mainly cardigan and long-sleeved jackets, long-tube skirts, waist skirts, short jackets with no collar and notches, long-sleeved jackets with Baotou and right ruffles, leggings, etc. The three-tailed snail hairpin is the most representative ornament.

(2) Shuangjiang style. It is mainly in Shuangjiang Lahu, Va, Bulang and Dai Autonomous County. Baotou and slanted right-breasted jacket are distinctive.

(3) Yunxian style. The old-fashioned clothing has its own characteristics, but now it has been changed. It is the same as the general clothing of the society.

(4) Simao style. It is similar to Shuidai clothing.

(5) Gengma style. It is same as local Dai costumes.

Costume of the De'angs

The De'ang people are mainly distributed in the Lancang Lahu Autonomous County of Dehong Dai and Jingpo Autonomous Prefecture, Baoshan City, Lincang City and Pu'er City. De'ang women wear black cloth baotou, and wear navy blue or black jackets with two fronts. They wear tube skirts, leggings, and waistbands. De'ang women's tube skirts are longer, covering the breasts at the top and reaching the ankle at the bottom. The skirt is woven with brightly colored horizontal stripes. Because of its different stripes and background colors, it is known as "Hua De'ang", "Red De'ang" and "Hei de'ang". On the skirt of "Hua De'ang" are red and black wide and thick horizontal stripes, and some are inlaid with four white stripes, and red cloth strips with a width of 17 cm are inserted into the withe stripses. The tube skirt of "Red De'ang" is woven with red horizontal stripes of about 17-20 cm wide. The tube skirt of "Hei De'ang" is woven with black thread, red, green, and white pinstripes woven on top. The length of the skirt is from the waist to the ankle, shorter than the skirts of the other two branches.

De'ang women wear five or six or even twenty or thirty waistbands around their waists. The waistbands are made of rattan or bamboo. Most of them are painted in red or black, and some are also engraved with various patterns. Most of them are put on the waist of the woman by the man when the man pursues the woman. As a sign of mutual love, the waist hoops are the characteristic decoration of De'ang women's clothing, and their cultural meaning is also very ancient.

Costume of the Dais

Dai costumes can be divided into three types: "Dai Le" (Shui Dai branch), "Dai Na (Early Dai branch) and "Dai Ya" (Hua Yao branch).

The "Dai Le" branch is mainly in Xishuangbanna. The women's clothing is characterized by elegance and purity. Girls wear hair buns on the top, slightly to the right, with flowers or colored hair combs on the bun. They wear a collarless sleeveless tights, which has a bra-like function. They wear a short jacket with a round neck and a right placket. The length is only over the navel. The colors of the shirts are mostly pure red, pink, water green, red yellow and so on. The barrel-shaped skirt is made of colored cloth, and the skirt length reaches the instep. They wear a silver belt, beautiful in width and fine lines. Belts are worn under blouses. They walk in bare feet or sandals. When they go out, they carry a "Tongpa" (a satchel) on shoulder. Women like to wear earrings, bracelets, necklaces etc.

"Dai Na" style clothing is mainly in Dehong and also seen in Baoshan, Gengma, Lincang and other areas. Girls' head is wrapped around the top with a red rope knot, gold and silver jewelry inserted. They wear a small hat made of bamboo. They wear a white or light blue right-breasted short jacket, with gold and silver dragon plaques or flowers on the chest, and black trousers instead of a barrel-shaped skirt. A small apron with cyan embroidery is attached, and a long streamer is tied around the waist and hangs on the right side, reaching to the ankle. Some also wear colorful shawls.

After marriage, women wear tall hats made of black cloth, with green headbands wrapped around the hats as decorations. They wear a short blouse and a black barrel-shaped skirt, wrapped in blue cloth from the knee to the ankle, which is convenient for work.

"Dai Ya" clothing is mainly distributed in Xinping and Yuanjiang. Women wear lace tight-chested short vests, which only reach the chest. The bosom is made of blue earth cloth or pink or grass green cloth, and a row of fine silver bubbles is nailed at the bottom. Women wear a black short shirt with long and narrow sleeves, silver bubbles on the collar, and lace on the body. They wear a black barrel-shaped skirt, which is thicker in texture, reaching between the knee and ankle, and the hem is decorated with 6 cm wide lace. The outer waist of the skirt has a total of three strips. From the inside to the outside, one layer is shorter than the other, showing each layer of lace, and the innermost layer is long to the knee. They wear a cross-stitched belt of about one meter long, embroidered with plants and geometric patterns. There are also two silver belts: one is belt-shaped, with larger silver bubbles inlaid with silver spikes at both ends, which are connected to the placket of the underwear to the lower left end of the front of the waist; the other is triangular and decorated with silver bubbles. Small, with a dangling thread down and up to the skirt, this belt covers the entire buttocks. When they go out, they hang a basket on the back waist. The basket is made of bamboo with fine workmanship and decorated with woolen flowers.

Costume of the Yis

The Yi nationality is the ethnic group with the most abundant clothing culture. Only in terms of their clothing styles, they can be divided into six types: Liangshan, western Yunnan, central Yunnan, southern Yunnan, northeastern Yunnan, and northwestern Yunnan. There are dozens of styles in each type, and each style has its own features.

Among them, the men of large and small Liangshan wear the headdress of "Heaven Bodhisattva" and the garb of "chaerwa". The cultural origin is very ancient. The clothing of women in central and southern Yunnan is extremely fine, which reflects the superb level of embroidery technology.

In Central Yunnan, taking Chuxiong area as an example, women wear short clothes, long pants, and a waistband. The waistband has two kinds of pocket shape and square shape, both of which are embroidered with exquisite patterns. Women's headwear is various, which can be generally divided into wrapping their heads with handkerchiefs and embroidered hats. The tops are mostly made of peach red, green and blue cloth, with few dark colors. Flower patterns are often embroidered on the neckline, shoulder support, cuffs and trouser legs. Some are also inlaid a circle of lace or decorated with colored silk tassels at the shoulder part, reflecting the colorful characteristics of the flowers more incisively and vividly. In particular, the satchel embroidered in the Tanhua mountain area of Dayao County has simple and elegant patterns and exquisite workmanship. It matches with men's and women's clothing and is more colorful.

In southern Yunnan, the Honghe area is the representative. Among them, "flower waist Yi" women's clothing in Shiping County is the most delicate and beautiful model in cross stitch embroidery. They wear a black or blue tight long sleeved large bodied shirt, which is knee long. The shoulder, hem and cuff are decorated with embroidery patterns and colored cloth. The outer cover is a unique pair of camisoles. The camisoles are almost made of fine cross stitch. There are also about 10cm cross stitch patterns and silver bubble inlays on the hem. The silver coin is a buckle. Women wear a top scarf cap, which is made of an embroidered towel with a length of 80 cm and a width of about 40 cm. Under the trousers, the back of the long shirt is often lifted up or folded over the waist, and the hem of the bottom is exposed. Embroidered belt has gorgeous colors and exquisite workmanship. The Yi women's clothing in Shiping is most unique in terms of the camisole and belt.

Costume of the Hanis

There are also many types of Hani costumes. Among them, the costumes of the three branches of "Aini", "Baihong" and "Yiche" are the most distinctive. Eni is concentrated in Xishuangbanna and Pu'er City. Women wear bamboo hats, called "Wuzhen", which are round and triangular. The hat is lined with black cloth. The front of the hat is wrapped with cream straw skin, and dozens of silver bubbles are nailed on the circle, and then surrounded by color beads and color lines. The girls wear small caps with silver bubbles and red feathers. Some are decorated with bone needles and insects made of cow horns.

Women wear black or cyan brassieres, about 30 cm long, with diamond shaped, round silver plates and several rows of small silver bubbles. The brassieres only cover the breast and are slung on the shoulder with a strap. The jacket has a round collar and a right placket. The back has a silk cross stitch pattern of triangular pattern, square pattern and wavy pattern. The cuffs are also inlaid with red, white and blue cloth blocks. They wear a short skirt that is not longer than the knee and is tied to the hip bone, covering only the lower half of the hip. Married people wear it lower. The waist is strung with a colorful belt of seashells. The two ends of the belt are hung with colorful beads, which is quite beautiful. The satchel is a woman's main wearing thing. It is made of white and black earth cloth, surrounded by flower cloth strips, embroidered with patterns in the middle, nailed with silver bubbles and silver coins, and then put on a ball of fluff, which makes the women's clothing of Aini more beautiful.

The women's clothing of Baihong branch is characterized by short, tight and small. The coat is not long enough for the navel. It is made of silver coins, but it is never fastened. It is beautiful to show the belly and navel. Six rows of silver bubbles are nailed on the front chest of the top, and an octagonal silver piece is in the middle, like a blooming white lotus. They wear the double fold shorts which is not longer than the knee and wrapped in embroidered leggings. Young girls wear small round hats with silver bubbles. Women wear blue cloth Baotou after marriage and childbirth. The whole set of clothing has the charm of modern women's bodybuilding.

Compared with other branches, the women's clothing of Yiche branch has stronger characteristics. Women's coats are divided into three layers, most of which are sewn with self-woven and self dyed homemade cloth. The innermost layer is a vest, the middle is a shirt, and the outer layer is a coat. There are 9 layers of outerwear, which are collectively called turtle clothes. The clothes are collarless, open chest, tight fitting, half cut. Not only the arms are exposed, but also the right chest is half exposed. Only the left chest is covered. Silver chain is worn on the front chest, which is matched with simple and elegant navy blue clothes, very elegant. An embroidered belt with a width of about 10cm is tightly tied around the waist. The front is decorated with silver coins and tassels, which not only outlines the curve of the girl's body, but also shows the girl's ability and heroism. A string of silver waist chains in the shape of more than 10 snails are tied outside the belt. It makes a jingling sound when walking. Women only wear a pair of blue shorts and are barefoot. From the sole of the foot to the root of the whole thigh, they are all exposed, even in the cold days. The shorts are very elegant in cutting. Generally, the girls sew them according to their own figure, and it is better to tighten the hips. The bottom edge of the lower end of the shorts should be herringbone, and seven folds should be folded in half, which looks like wearing seven overlapping shorts. The most conspicuous thing is that every Yiche woman likes to carry a white cloth umbrella with her, which is very conspicuous against black clothes.

The headgear of the Yiche woman is a triangular white pointed headcloth, shaped like a rain cap on a raincoat, but with a pair of beautiful swallowtails on the back, and the edge of the scarf is embroidered with exquisite colorful patterns. This kind of headgear is related to ancient legends and has a special cultural connotation of clothing.

Costume of the Bais

Women's clothing has small collar or collarless right Lapel underwear which is long enough to reach the crotch or knee. Young people mostly wear white and light blue, and middle-aged and above tend to wear blue and black. The outer cover is red or light blue cape shoulder; the right Lapel knot is hung with "three whiskers" or "five whiskers" silver ornaments; the waist is embroidered Square waist; the hips are hung with streamers. The

Bai women wear the light color wide legged pants and embroidered shoes. Their head is wrapped with embroidered or printed square handkerchiefs and colored towels. Unmarried women weave a single dish on the top, and married women wear a bun instead. Unmarried Bai girls in Jianchuan wear small hats Or "drum nail cap" or "fish tail cap" covered with jade rabbit silver bubbles.

Men usually wear white or light blue tops, black cloth or white sheepskin collars, and wide-legged, embroidered sandals with tassels or scissors buckle shoes. They like to wear white or light blue Baotou, hanging down about a foot. Fishermen by the Erhai Sea like to wear small caps with melon skins and tops with many buttons and short enough to show the navel. The more layers, the better-looking. They wear red and green belts around the waist, a purse, and a suede or cloth apron. There are many layers of clothes, which is called "thousand-layer lotus leaf". Wearing three layers of inner and outer shorts is called "Sandieshui", which is regarded as handsome and rich.

Costume of the Naxis

Naxi women wear right large-bodied fat sleeved clothes. They wear a collarless jacket with short front and long back. The waist is wide and the sleeves are large. The sleeves are as long as the arms, and the cuffs are rolled out about 10cm. A dark red or dark blue woolen collar coat is usually worn outside the top, which is also called "Camisole". Those who wear more than 3 camisoles can have a narrower shoulder from the inside to the outside, which means that they are rich. They wear long pants with colorful ribbons, pointed embroidered shoes and a pleated long apron. When not in use, it should be folded and stored to keep the pleats flat and tidy. "Seven Star cloak" is the most distinctive dress of the Naxi people. It is also called "Pi Bei" because it is worn behind women. The production is very exquisite and rich in cultural connotation.

Costume of the Lisus

Lisu People are distributed in Nujiang Lisu Autonomous Prefecture, Diqing Tibetan Autonomous Prefecture, Dehong Dai Jingpo Autonomous Prefecture, Baoshan City, Lincang City, Lijiang City, Dali Bai Autonomous Prefecture and Chuxiong Yi Autonomous Prefecture and so on.

According to the clothing, the Lisu people are divided into three branches: "Black Lisu",

"White Lisu" and "Flower Lisu", each with its own characteristics. Hualisu women in the Yingjiang area wear Baotou handkerchiefs.Yellow and white cloth strips are interlaced and inlaid, and small silver bubble lace about 20 cm is nailed. The headband is decorated with red spikes at one end. When Baotou is wrapped, women place the red end on the left side, hold the left hand down and wrap the other end from right to left for three circles, and then tie it tightly. Then, they put a red, yellow and white scarf on the tied handkerchief, embroidered with arrow patterns, hanging colored balls and red spikes. The top is a blue cloth long shirt, covered with a camisole made of colored thread, and they wear a two-layer apron. The inner layer of the apron is long. The lower end is embroidered with flowers and geometric patterns, and the outer layer is short. It is as long as the front of the abdomen, bordered with red, yellow and white cloth strips. The lower end is inlaid with seashells and embroidered with geometric ribbons. Lisu people wear a rattan ring on the calf and hang a mouthstring (musical instrument) on the chest.

Costume of the Lahus

There are mainly two types of Lahu women's clothing: long clothes and trousers and short clothes and skirts, which are characterized by long clothes. The long coat is a black cloth shirt with long sleeves on the right side. The two sides are slit at the waist level. The color strips are combined into geometric shapes along the edge, and then the silver bubbles are arranged into staggered large and small triangular patterns, which are regularly embedded in the high collar, chest circumference and sleeves. The edges of the clothes are sewn with wavy ribbons or triangular and rectangular patterns of colored cloth, and also with diamond ribbons embroidered with white wavy patterns. In addition, they are carefully inlaid with silver ornaments, giving people a feeling of brilliance, beauty and luxury. They wear black cloth pants and wrap their heads around with a ten meter long Baotou. Baotou is available in black and white, with colorful tassel dangling at both ends. One end of the adult's Baotou hangs down to the back of the waist. They wear leggings, barefoot. Men and women have the custom of shaving their heads, but women leave a strand of hair on their heads, wearing large earrings, bracelets, and silver medals on their chests.

Costume of the Jinos

Most of the Jino people live in Jinuo mountain, Jinghong City, and a few in mengwang

Township, Jinghong City.

Men wear long sleeved short clothes with no collar and no buttons, which are tied with cloth strips. The clothes are long to the navel, and are mostly sewn with linen, with red, black and yellow interlaced stripes woven on a white background. There is a 20cm black embroidered Square on the top of the garment. The square is embroidered with black edges, red, yellow and other radioactive circular patterns are embroidered in the middle. Colored stripes are embroidered around. This square pattern, called "Moon Flower", is the symbol of Jino men's clothing. Men like long cloth Baotou. Men wear pants and earrings. The bigger the hole, the more beautiful it is.

Women wear long open linen short clothes without collar. The body is made of striped cloth with red, black, green, yellow and calico. The middle part of the back is inlaid with a red or colored cloth, and the two shoulders are also decorated with a lace. The front of the garment is open, and a breast guard handkerchief made of colored cloth or with patterns is tied inside. There is also a pocket shaped chest handkerchief inlaid with silver bubbles between the two lapels to cover the front abdomen or the lower chest. The lapel is unbuttoned and fastened with small cloth strips. The skirt is short to the knee and has a single piece. It is made of colored cloth or striped linen. The waist of the skirt is inlaid with a piece of linen woven with stripes, and the laps on both sides of the skirt are inlaid with lace. They wrap black cloth leg cover under the knee. Women wear a pointed hat, which is made of five color striped linen. The pointed hat is like the wings of a bird. It was created by Jino girls imitating the shape of a bird.

Jino women like to decorate their heads with flowers, and their heads are decorated with five colored tassels. They wear earrings, bracelets, collars and rattan rings made of thin bamboo strips on the waist.

Costume of the Dulongs

Dulong nationality, with a population of about 7000 (census data in 2010), is mainly distributed in Gongshan Dulong and Nu Autonomous County, Nujiang Prefecture, Yunnan Province.

The traditional costumes of the Dulong people are relatively simple, but very distinctive. Both men and women bare their arms, with only one or two self-woven linen blankets draped diagonally. The linen blanket is flanked from the armpit at one end to the shoulder at the other end. Both men and women wear linen leggings on their lower legs. Underage boys use rattan to hang a small basket over their genitals. Adult women have the custom of tattooing face. Men break their hair with a machete, They hang a sharp blade on the left and a bamboo basket on the right.

After the 1960s, great changes have taken place in Dulong clothing. Women have worn the common clothes and skirts in the mainland, wearing earrings and pendants, braiding top scarves and so on. Men's clothing is also mostly modern clothing, but a linen blanket is added on the outer garment to reflect the traditional national characteristics.

Special clothing traditions have created special clothing craftsmanship and dressing customs. The traditional costume of the Dulong nationality - linen blanket, also known as "Dulong blanket", has not only become the unique handicraft of the nation, but also the characteristic symbol of the nation.

Costume of the Jingpos

The Jingpo people have a unique style of clothing. The main colors of men's clothing are black and white, and the old men's clothing is the same in all branches. As for middle-aged and young men's clothing, there are subtle differences between the Jingpo branch and other branches. They wear a white stand-up collar shirt, a black round-neck coat, black trousers, and a red and blue checkered cotton gauze circle on the head. The middle-aged and young men of several other branches all wear white shirts, black trousers, and white toe caps decorated with pompoms and tassels of various colors. No matter which branch of the men travels, they will carry a handkerchief (that is, a backpack) and a long knife with him.

Women's clothing is divided into casual and dressy. If wearing casual clothes, the upper body should be black or various colored tights with the front or the right front, and the lower body should be a long cotton skirt of solid color or woven with the characteristic patterns of the Jingpo people. Dress is the dress for festivals or weddings. The top is a black short-breasted shirt with no collar and narrow sleeves. The chest, shoulders and back are decorated with silver bubbles, silver medals and silver spikes. Women wear a wool barrel-shaped skirt with a red belt around the waist.

The headdress is a red bottom flower bag made of wool. The lower leg is wrapped with a leg wrap of the same texture and color as the skirt, and several strings of red beads, earrings and bracelets are worn.

Women wear a cylindrical hat made of brocade. They wear a black round necked narrow short coat with hundreds of silver bubbles, silver medals and silver spikes inlaid on the front and back of the shoulders. A collar, a silver chain, a silver bell, etc. are hung around the neck.When they dance, they will chatter. Women wear silver ear tubes, some longer than fingers. They wear an engraved silver bracelet. Women are the most beautiful in wearing silver.Some women also wear rattan waist bands in red or black paint.

Men wear a turban. One end of the hood hangs down, and it is decorated with pompoms in bright and eye-catching colors. When going out, they hang a "Tongpa" (a shoulder bag) and a long knife on the waist. The tube handkerchief is red, decorated with neat silver bubbles and silver tabs, and a long tassel below. The red bags with white knives show the valiant attitude of Jingpo men.

Costume of the Jingpos

The Miao ethnic group has a wide distribution, many branches, and complex costumes. The headgear, clothing and skirts have their own characteristics and splendor.

(1) Zhaotong Huamiao. Also known as Dahuamiao, it is found in Zhaotong, Kunming, Qujing, Chuxiong and other places. Women's traditional pointed bun headdress and flower shawl are the most distinctive. The pointed bun headdress has been changed a lot, and the shawls embroidered with legends and historical content are generally preserved.

(2) Baimiao. They are distributed in Zhaotong, Wenshan, Honghe and other places. They are all called Baimiao. Clothing is very different from place to place. It can be divided into five types: Zhaotong, Jinping, Funing, Xichou and Malipo. The difference between the headgear is the most obvious. Except for the Zhaotong style, the clothes are basically similar, with more open-breasted shorts, pleated skirts, and more streamers. Zhaotong Baimiao also wears a pleated skirt, with a slanted front, and a thin disc wrapped around the head. The Jinping-style headdress is a wide Baotou, with one end hanging from the right chest and the other hanging from the forehead. Funing style is a barrel-shaped Baotou. Malipo style is the long Baotou into stacks of herringbone seams. Xichou is an inverted bowl-shaped Baotou with an octagonal pattern embroidered on the flat top. Xichou Baimiao

still retains the embroidered traditional men's long shirt.

(3) Jinping Heimiao. It is in the name of clothing color. Baotou is a long and round barrel. They wear a slanted skirt.

(4) Qingmiao. It is distributed in Guangnan, Funing Wenshan, Xichou, Malipo and other counties, and there are differences in clothing. Funing Qingmiao wears a big black braid, a small coat and a long skirt.The pleated skirt is long to the ankle, and the dress is tight and small. There are also Qingmiao in Lufeng County. They wear headbands and trousers. The front is short and the back is long.

(5) Hongmiao. There are some differences in different places. Hair buns and slanted-breasted clothes are common. Funing Hongmiao has changed into modern fashion.

(6) Lvmiao. It is in the name of clothing color. They wear a skirt with lapels.

(7) Huamiao. There is Huamiao in Wenshan City, Qiubei County, Maguan County, Malipo County, Mengzi City, Weishan Yi and Hui Autonomous County and Huaning County. Wenshan Huamiao people wear a collar cardigan, long apron with front and back streamers and pleated skirts as the main clothing accessories. They wear thin and neatly stacked round head plates. In some places, beads and silk whiskers are dropped on the edge of the head plate. It varies from place to place.

(8) Jinping Tongchang type. It is distributed in Chaoyang and Tongchang of Jinping Miao, Yao and Dai Autonomous County. Red Baotou, red belt, square apron, skirt and collarless round neck slant Lapel are typical clothing.

(9) Jiaobiao type. It is in Kaiyuan city. The main feature is a long braided headdress that is folded like a double angle.

(10) Hanmiao. It is distributed in Wenshan Zhuang and Miao Autonomous Prefecture and Huaning, Jinping, Shizong, Mengzi and other counties. The costumes vary from place to place. The Jinping area is also known as Piantoumiao.

(11) Qingshuimiao. It is distributed in Jinping Miao, Yao and Dai Autonomous County.

(12) Ninglang style. It can be found in the Ning Yi Autonomous County. Baotou is overlapped with herringbone seams. People wear a long coat with a slanted skirt and an apron or pants under the skirt.

Chapter Three   Hair Accessories and Headwear

In the process of decorating their heads, all ethnic groups in Yunnan can creatively display their artistic talents mainly in hairstyles, hair accessories, crown accessories, ear and face accessories. Although the head is not big, the position is important and can show the decorative effect, so people pay attention to decorating the head. There are many types of headgear, which are highly artistic and exquisitely set, and play an excellent embellishment role in the whole dress collocation.

From ancient times, the hairstyles of Yunnan ethnic groups can be categorized into buns, braids, shaved hair and bald heads, each of which has many subtle differences. Hairstyles and hairstyles are all embodied in women. Except for a few Yi people in Shizong County who have long hair, men all follow the popular habits of modern society and rarely have their own characteristics. Among the De'ang ethnic groups, women from the "Bielie" and "Liang" branches shave their heads. Lahu women in Shuangjiang Lahu, Wa, Bulang and Dai Autonomous County also shave their heads after marriage, which is a unique hairstyle.

The three hairstyles of draping, braiding, and bun have been used in character modeling as early as the Qin and Han dynasties. At present, these three hairstyles are still mostly different in shape.

There are not many specially made fixed decorative hairdos, most of which are the relaxed and leisure state of braids and buns. To drape the hair is to drape the loose long hair behind the head and let it loose without other decorations. It can also be simply wrapped with something like a headscarf, or it can be tightened with a rope.

There are many kinds of braids. Long hair is not too short, but it can not meet the requirements of braiding. It is often necessary to connect some things. The common method is to connect wig braids and wiring braids. Wig braids are the collection of long hair that falls when brushing and braided for later use. The braids are made of large braids with threads, which are coiled on the head for decoration. In order to be in harmony with the braids, black threads are often used. Occasionally, some people use colored threads to highlight their decorative effect. Pumi, Tibetan, Hani, Miao, Yi and other ethnic groups all use thread braids. Naxi Mosuo people also use thread braids, and some ethnic groups such as Yi and Mongolian use hair braids. Due to the differences in local habits and marital status, there are many different braiding methods such as single, double and multiple braids, and the way of braiding is also different.

The braids should be tied tightly with a head rope. Most of the braids are wrapped with gauze, scarf and headband, and they are arranged into a certain hairstyle, and there are not many exposed hair.

The most famous hair bun in the history of Yunnan is the cone-shaped bun. People refer to the cone-shaped bun in different periods in different periods.

In Qin and Han Dynasties, the cone-shaped bun is a spindle-shaped bun, tied tightly at the mid-waist, large at both ends, and dragged at the back of the neck. The so-called cone-shaped bun in later generations mostly refers to the top of the head. The buns are all wrapped in various head accessories, and are rarely exposed. They are rarely seen and are not very important. Most of them are to create a head shape and make a simple bun, which is relatively casual, but there are also many fixed hairstyles. Common buns include top bun, drag bun, loose plate, and bowl-shaped bun. The top bun includes spiral bun, pointed bun, lotus bun-shaped bun, columnar bun and so on. The hair is casual, and the naming cannot be very accurate.

There are many kinds of accessories on the hair, such as Zan hairpins, Chai hairpins, hairpins, ornaments, etc. Some are used to fix the hairstyle, and some are used for decoration.

Zan Hairpins are commonly used to fix hairstyles and also play a decorative role. As early as the Qin and Han dynasties, various ethnic groups in Yunnan have used them in large quantities. Zan hairpin is a long pointed body, which is inserted into the bun to fix it. The main pattern appears at the tail of the hairpin. Most of them are made of flowers, birds, insects and fish. In strips, some are made into birds, butterflies and other shapes, and there are many styles. In addition to gold and silver, the materials used are also jade, bone teeth, bamboo, and wood.

The function of the Chai hairpin is similar to that of the hairpin. The hairpin is double stranded, and it is also painted and carved at the tail. Some also hang chains and pendants for decoration. Usually deep into the hair, the hairpin can be higher than the hair ornament, and the hanging ornament has a unique decorative effect. The head hairpin of the Yao nationality is like this. It is inserted into the hair bun, and the tail of the hairpin and the ornaments are exposed.

Adding a special bamboo tube on the top of the head to tie it with the hair coil and wrap it around the head to set the shape is a method of decorating the head of some Hani people in Xishuangbanna. The bamboo tube is a unique ornament.

Yao people use a silver sun pattern round top plate, which is also a round head plate, in which the silver bar is connected with the middle as the skeleton, and some are made of new leaves, banana leaves and bamboo shoot shells for convenience.

Lezi is an ornament of many nationalities. It is decorated on the upper part of the forehead and hairline. It is a curved semicircular bar, carved with patterns and patterns, or decorated with breast nails. There are also pendant chains for decoration, and some are made of cloth or other articles, decorated with beads, jade, silk whiskers and other things. It is also called Toutiao.

The decorative comb is also a hair ornament. There are silver products and other products. Some are directly inserted into the hair, and some are connected with the chain. The headband is famous for Va people. It is not only used for decoration, but also for fixing hair styles. The silver headband of Hongtou Yao belongs to another type.

To braid or bun, people should wrap your hair with gauze, silk screen, headcloth, handkerchief and hat. In the past, the hair was covered with black scarves and silk scarves, and then the head scarf or hat was wrapped, or only the scarves could be wrapped. At present, the popular flower scarves are all woven square scarves.

Baotou and handkerchief are much the same, and can not be distinguished by the shape of cloth. Most of them are rectangular cloth, and there are also short handkerchiefs, which are wrapped into a certain shape and fixed for use. They are also called headgear.

According to the shape difference when wrapped and placed on the head, Baotou of Yunnan nationalities can be classified into about 30 kinds.

(1) High barrel-shaped Baotou. It is as round as a column, higher than the diameter, and shaped like a barrel. It may be that the upper part is slightly smaller than the lower part. Some are also decorated with ornaments, some are wrapped with cloth, and some are fixed. This kind of Baotou is found in the Liangshan branch of Yuanmou County, the Yi people in Shizong and other places, the Zhuang people, the Jingpo people, the Huayao Dai people, and the Nujiang Lisu men's clothing in Yuanyang County.

(2) Flat barrel-shaped Baotou. It is also cylindrical, with a diameter greater than the height, and is externally decorated with tassels. Hongtou Yao in Jinping, Hekou and other places have this kind of Baotou, and Hualuo in Yi people in Funing County has this kind of Baotou, with different thickness.

(3) Big wheel-shaped Baotou. The small cloth bands are stacked layer by layer and wrapped like a wheel, leaving a hollow part on the head. The thickness is less than a few centimeters, and the diameter can be large or small. It can be divided into several types. The edges are also decorated with fine ornaments such as silk whiskers and beads. Miao people in Wenshan, Dayao Yi people, Xinping Hani Kado people all have this kind of headwear. Kado people also use cross cloth belts to hold their heads and wear silver bubbles.

(4) Twisted wrapped Baotou. It is used by both men and women. The overall shape is round, and the head is hollow. It is not required to be neatly wrapped to form a twist. Yi people in Luquan Wuding and other places also wrap this kind of headdress well and set it. When it is used, it is covered with silver ornaments such as flowers and butterflies.

(5) Herringbone Baotou. The main feature is that the Baotou at the front part is folded into a zigzag shape. Miao, Lisu and other nationalities have this wrapping method.

(6) Short handkerchief Baotou. There are many types. Some use short square cloth, and some use towels to cover the head. Sometimes, the two ends of the handkerchief are decorated with patterns, embroidered, or decorated with Tassels and beads, which are exposed for decoration. Towel Baotou is often matched with casual clothes and is used in many places.

(7) Headgear. The outside is made of hard cloth, and the hollow part is covered with

cloth, which can also expose the hair, and follow the custom. Yi Sani people in Shilin, Luxi, Maitreya and other places, and the Hani people in Shiping County have such headgear. Because of their thickness, length and color, they form different types.

(8) Jiaojiao Baotou. The tail end of Baotou is deliberately left up to form a unique headdress. The Nong branch of the Zhuang nationality and the Bulang Nationality are typical.

(9) Fuwa shaped Baotou. The appearance of Baotou is like a combination of two plates. For example, Dai nationality in Maguan County, Dai nationality in Shiping, Yuanjiang and other places, and the Shayao branch of Yao nationality in Jinping Miao, Yao and Dai Autonomous County all have this kind of headdress. In some places, it is the dress of unmarried women.

(10) Inverted basket-shaped Baotou. Baotou is like a flat column, and the upper part is slightly small, such as a back basket. Dai people in Shiping, Yuanjiang and other places have this kind of Baotou.

(11) Inverted bowl-shaped Baotou. It is shaped like an inverted bowl and covers the head. It is round and large at the top and small at the bottom. Miao people and Dai people in Jinggu and other places have such headwear.

(12) Baotou with horns. Sharp corners are at the front end or other parts of Baotou, including one corner and many corners. Hani people in Yangcha street of Yuanjiang Hani, Yi and Dai Autonomous County, Lisu People in Baoshan, and Bai Yi people in Shilin Yi Autonomous County all have such headwear.

(13) Square Scarf Baotou. A Square towel is a new type of woven product, which was popular in many places in the 1980s and before. It is usually wrapped in braids and buns.

(14) Dali Bai headwear. It is mainly popular in Dali City, Eryuan County and other places. Bai nationality in Lijiang Naxi Autonomous County and other places also have this headdress.

(15) Landiaoyao headdress. The sun pattern round top plate, the chain pendant and the connected columnar bun are the main components.

(16) Xichou Hualuo Baotou. It is popular in Jijie township of Xichou County and some neighboring areas.

(17) Xiaoliangshan Yi style bun. It is different before and after marriage. the plate-shaped bun is the main symbol.

(18) Folding Baotou. Among Zhuang Tuliao, Datou Tuliao, Pingbian Yi and Weishan Xishan Yi all have this wrapping method. The main feature is the folding of cloth strips.

(19) Pointed Baotou. The headdress is pointed and protruding, some are covered with cloth, some are wrapped with cloth, and some have a large number of silver ornaments and long ribbons. Jiantou Tuliao and Heisha people in Zhuang nationality, Jiantou Akha in the Hani nationality, Hongtou Yao in Jinping and Hekou all have such Baotou. Their appearance is very different from each other. Hongtou Yao is the dress of married women.

(20) Changyi Baotou. It is mainly popular in Lancang Lahu Autonomous County and some places in Xishuangbanna. Baotou is a coat and a special headdress of Hani.

(21) Nanmei Lahu Baotou. Mainly in the area of Nanmei Township, Lincang City.

(22) Yi Huayao Baotou. Also known as hat, in Shiping County and Eshan Yi Autonomous County, it is complicated to make and wear. The flower and bird patterns on the back are very eye-catching and the decorative effect is excellent.

(23) Draped Baotou. After Baotou is wrapped, the two ends are exposed from the lower part of the inner side and hang down, becoming a decoration.

(24) Daheiyi style Baotou. Popular in Shilin, Luxi, Maitreya and other places in Dahei Yi, it is a unique wrapping method.

(25) Baotou with a cover. Covering the head of the Baotou for decoration, Yi, Hani, Jino and other ethnic groups have such headgear.

(26) Flat front and back inclined Baotou. Zhuang nationality flat-headed Tuliao and Muji people of the Yi nationality have this kind of Baotou.

The above are just some common Baotou methods. There are many Baotou styles, and the changes are also random. It is difficult to list all.

Besides Baotou, the hat is also an important cover and ornament for the head. Hats are available to men, women, old and young. Children's hats have the most shapes and styles. Some children's hats are similar to those of adults, and more are unique. Goldfish shaped cap, tiger head shaped cap, lion head shaped cap, peacock cap, cat ear cap, flower and bird shaped cap, magpie cap, archway cap, whale head shaped cap, dome shaped cap, rabbit ear cap, lotus cap, Narcissus cap and many other children's caps are common and not limited to a single nation. Hani children's hat, Yao children's hat, Yi chicken hat, Dai children's hat in Xinping, Yuanjiang and other places in Xishuangbanna are unique hats with national characteristics.

Adult men's hat is the most characteristic of the Yao's horsetail hat, Naxi Dongba five Buddha crown, Jingpo's brain double hat, Tibetan leather hat and Hui's white hat.

The most representative are women's pointed hats of Jinuo, Menglian Fuyan Wa and Hani Yiche people, Luoguo shaped hats of Guihua Yi people in Dayao County, Luoguo shaped hats of Yi people in white clothes in Heqing County, peacock open screen shaped hats of Yi people in Pingbian Miao Autonomous County, beaded net hats of Lisu People in Fugong County, flower hats and cockscomb hats of Yi people in Yuanmou, Wuding and other places, Casa Dai girl hats of Xinping Yi and Dai Autonomous County, Luquan Wuding Yi wool hat.

Cockscomb hat is a very popular hat among the Yi people. It can be roughly divided into Honghe style, Yuanyang style, Lvchun style, Awu style, Mengzi style, Chuxiong style, Lufeng Hongyi style, Kunming Xishan style, etc. Chuxiong style is also called the parrot hat. Lufeng red Yi style is also called the butterfly hat, and the magpie hat of the Yi people in Tonghai County is also a variant. In addition, Bai people also have a cockscomb hat similar to the Xishan style, called the Phoenix hat. Hani people in Leyu Township, Honghe County have the same cockscomb hats as the Yi people.

Modern brim hats, military hats, etc. are also very popular in many places, and are worn by both men and women. Due to the limitation of fabrics, military caps were rare when they appeared. They were rare in the 1960s and 1970s. Many were proud to have one. Since then, they have been popular and have continued to today.

On the hats and Baotou, there are often ornaments. The hats are decorated with silver medals carved with Buddha statues and the characters of longevity and happiness. There are also ornaments such as fish dragons. Baotou has more ornaments, such as silver medals and silver chains, and is decorated with Tassels and silk whiskers. Jade, agate, animal claws, animal teeth, seashells, material beads, etc. are also commonly used as head ornaments. The tail feathers of long tailed birds such as golden pheasants and pheasants are also used for head decoration.

Earrings are also an important part of headwear. Commonly used earrings among ethnic groups in Yunnan include Erdang earrings, earrings and Erzhui earrings. Erdang Earrings are also called ear posts, which are directly inserted into the perforations on the earlobes for decoration. The more common earrings include cylindrical ear posts, mushroom-shaped ear posts, cylindrical ornamental chrysanthemums, flower-shaped head ear posts, lotus-shaped silver-clad jade ear posts, and long-rod-shaped ear posts decorated with tassels and chains. Ear posts are not used in many ethnic groups, and it is mainly used in the De'ang, Dai, Blang, Wa and other ethnic groups.

The use of earrings and eardrops is very common, almost every ethnic group has them. There are many styles of earrings, usually they can be made into thin circles of gold and silver earrings, single-strand head and tail engraved earrings, engraved gold and silver earrings, and jade earrings. Single-strand head and tail engravings are usually made into spring shape, fern bud shape and flower shape. The engraved patterns on the earrings are mostly flowers, birds, insects, fish, plants and the like.

Eardrops are mostly composed of earrings and pendants. The rings are mostly made of gold and silver coils, and there are also more complex shapes. The part of the pendant is more complicated, and can be made into leaves, fruit flowers, etc., and can also be made into other natural objects and artificial products. There are also a large number of pendants and animal shapes.

Chapter Four   Limb Decoration

Limb ornaments mainly include clothing patterns and ornaments on the limbs. Patterns mainly appear on the sleeves, trousers, skirts, leggings, and shoes of tops, which not only emphasize the coordination of the whole body, but also have their own characteristics. Ornaments include finger, wrist, arm, foot, ankle and leg.

The part of the skirt that hangs down to the lower limbs is the main load-bearing part of the patterns and patterns. The patterns are usually made by color matching, pleating, weaving, and dyeing. The batikers include Miao people and Yi people from Malipo Xinzhai. Jingpo people are the most exquisite in weaving, and there are many people who make pleats. Some ethnic groups also embroider lace on the edge of the skirt for decoration.

The pattern decoration of the trousers can be divided into several types, some are decorated with red, yellow and white threads in the middle of the trousers, some are decorated with lace or embroidery on the edge of the trousers, and some are decorated with geometric embroidery on the whole trousers. pattern or flower and bird pattern. The edge of the trousers is decorated with lace, which was used in women's clothing in many places in the past, and the embroidery is found in a few ethnic groups such as the Lahu people. There is a picture of the Lahu people embroidered with sun patterns. The embroidered geometric patterns are typical of the Yao nationality in Jinping, Mengla and other places. They are embroidered with tree-shaped group diagrams, Wan (swastika)-shaped group diagrams and other various geometric pattern puzzles, and are coordinated with various colors such as red, yellow, blue and purple. Pattern, resulting in a striking dominant color. There are also many kinds of costumes with full trousers embroidery. Taking the costumes of the Yi nationality in Yongren County as an example, the composition materials are flowers and birds, and the combination of lace and embroidery patterns, plain and color matching methods are used to embroider exquisitely coordinated trousers decoration. The Yi people in Funing County use a triangular mosaic composition, and use red, black, yellow, white and other tones to make beautiful patterns.

There are three types of leggings: long strips, angled strips and sleeves. The decoration of the long cloth belt is mainly decorated with flower spikes on the edge, and it can also be decorated with lines on the edge, which can be wrapped into flower lines when tying. The ornaments are bayberry flowers, mostly for men. Both angled pieces and sleeves can be embroidered, but there are those who embroider and those who don't. The beautifully prepared leggings also have excellent decorative effect on the calf, often with the feeling of trousers under the skirt or pants hidden under the big trousers, and some clothing with exposed knees, pants, skirts, leggings, and unique decorative effects.

Sleeve composition includes large and small sleeves, single-sleeve and cloth coloring, color coil decoration, forearm embroidery, cuff embroidery, mid-sleeve embroidery and full embroidery. In some garments, the sleeves are separate from the clothes, and they are put on and fastened together to form one body. The independent big sleeves of Zhaotong Miao people are an example. Large and small sleeves are mainly used in the popular sister clothes and half-cut Guanyin clothes in the past. The sleeves of these clothes are short in the large sleeves and long in the small sleeves. Embroidery has a unique decorative effect. Among Zhuang, Buyi, Yi, Mongolian and other ethnic groups, there are still people who keep this kind of clothes.

For the sleeves with the same fabric and different colors, some use the same fabric and different colors to create the effect, and some use the method of connecting the cloth with the brocade to create a special decorative effect. Hani, Zhuang, Yi and many other ethnic groups use this method.

Coil decoration mainly installs lines of different colors such as red, yellow, white and blue on the sleeves to form coils, and the decorative effect is highlighted by color matching. Jino, Lahu, etc. all use this method to decorate sleeves, and sometimes small cloth strips are used to make coils.

Whether it is forearm embroidery, mid-sleeve embroidery, cuff embroidery or full-sleeve embroidery, the subject matter is mostly patterns such as butterflies, flowers and birds. The patterns on the forearm and the whole sleeve are mostly connected in a circle, layered from top to bottom, and sometimes decorated with lace. The batik sleeve patterns of the Yi people in Xinzhai, Malipo County, and the thick geometric patterns of Miao people in Zhaotong and other places are two special ones. Others add floral cloth or yellow, white and red coils for decoration on the cuffs.

Shoes are further divided into shoes and boots. Boots include tube shoes. Tibetan leather boots are the most used. In some places, cloth boots are also preserved. There are tube embroidery, which is rarely done at present. There are many types of shoes, which are divided into materials, such as straw sandals and cloth shoes. Cloth shoes are divided into general shallow shoes and cloth sandals. The clogs, bark shoes, and bamboo-straw shoes used by the Hani and other ethnic groups are daily necessities with special materials and shapes.

The shape of cloth shoes, in addition to the shape of the upper and the bottom, the pattern of the upper and the pattern of the toe are different. According to the pattern of the toe, there are spotted shoes, cat-toe shoes, fish-shaped shoes, zodiac shoes and so on. In terms of the shape of the sole and body, boat-shaped shoes are special. The embroidery on the upper is mostly patterned with peony, butterflies, flowers, birds, insects and fish, as well as combination patterns with auspicious meanings such as mice, grapes, peach and bergamot.

Finger accessories are mainly rings, finger rings and finger hoops, most of which are simple rings, made of gold, silver, jade and other materials. Some of the fingersstalls are the expansion of the finger ring. The gold and silver finger rings are stacked like a spring coil, which can be called a finger hoop. There are also ring sleeves, which can be decorated with flowers or plain, and some fingerstalls are shaped like a big arm hoop, but are reduced to fit on the fingers.

There are quite a lot of shapes of rings, and they can often be made at will, including two parts, the ring and the ring surface, mainly on the ring surface. The ring face has round, square and other shapes, mostly square and round, but also shield, hexagon, animal or other geometric patterns. The ring face has a variety of compositions such as plain face, lettering, coin face, and engraving. The engraver can engrave a single word, and some make a seal. The engravers have single engraved flowers and birds, and can also engrave festive pictures such as magpies and plums. Some are also decorated with jewels and jade, which can be single or multiple. Some ring faces are also hollowed out for drawing, and some rings are also decorated with small pendants.

Wrist accessories mainly include bracelets, chain bracelets and hand rings. Bracelets have different textures such as jade, emerald, amber and gold, silver, copper, iron, etc. There are straw bracelets, which are used for children. Jade bracelets are mostly made into plain round surface, without engraving patterns, winning with their own color and material, and occasionally covered with silver and gold. Iron bracelets are only occasionally seen in a few places such as Dulongjiang, and there are not many. Due to the low price, the processing of copper bracelets is not very particular, and the ring twisted silk bracelets are more common.

The processing of gold bracelets and silver bracelets is much more complicated. There are many methods such as hollowing out, twisting, carving, shaping, female flowers and so on. The common bracelet body shapes include square thin circle, round, hexagonal, drum, flat circle, twisted wire, wound wire, and so on. Some bracelets are made into a circular ring and inserted from the fingers, while others have openings and are inserted from the wrist. The designs of the bracelet are mostly flowers, birds, insects, fish, auspicious animals and dragons. Carved patterns are mostly carved, butterflies, trees, leaves and so on. There are also canines, ripples and other geometric patterns. The flowers are decorated with auspicious animal patterns such as lions and unicorns, and the bracelet heads are matched with dragon heads, dragon playing beads, and dragon armature beads. Some bracelets are decorated with pendants.

The shape of bracelets varies not only from place to place, but also from gender and age, mainly reflected in the size, thickness and pattern. There are also differences in wearing methods. Some can wear one, and some can wear five or six or more at the same time, extending from the wrist to the forearm. Traditional bracelets are mostly used by children. There is a long-life lock on the chain, which is not only used as decoration but also as an amulet.

The ornaments on the arm include arm bracelet and arm band. The arm bracelet is close to the bracelet, but the circle is larger. The arm hoops are usually made into long or short tubes, with curled mouth. The hoop body is decorated with carved or inlaid lacquer. There are also tube openings and spring coils. The length of the long arm band is about nine cm, or longer. Not many nationalities use the arm band. Va, Jingpo, Hani and other nationalities use them for the small arm. Even the arm band on the wrist is also classified as a bracelet.

Ankle ornaments are mainly foot chains, belonging to children's ornaments. They are used for body protection and decoration. Some ethnic groups such as Va have leg bands made of bamboo and cane.

Chapter Five   Accessories

Accessories are some independent individuals, most of which have certain functions and are tools. At the same time, they become the personal belongings used by a certain nation or some people, and play a decorative role. They are special members of the clothing system, and some are pure decorations.

The common accessories of Yunnan ethnic people include sachets, pendants, hanging boxes, areca bags, hydrangeas, amulets, bibs, back covers, baskets, straw hats, bamboo hats, umbrellas, coir raincoats, satchels, money bags, chains, knives, crossbows, etc.

Bai and Yi people in Dali have sachets, which contain spices or grains. They can be hung for decoration and can also be worn. In the past, they thought they had the ability to ward off evil. The common shapes include people, monkeys, tigers, butterflies and other animals, as well as Buddha's hands, pomegranates, flower baskets and other shapes. The body of the bag is also embroidered with various auspicious patterns and auspicious characters such as happy birthday. There are also small pendants for similar purposes.

Some small hydrangeas, like sachets, can be worn. Some ethnic groups, such as Zhuang, Bai and Dai, have hydrangeas, and their workmanship is more exquisite.

Areca bag is a small decorative bag used by Yao people in some areas of Wenshan Zhuang and Miao Autonomous Prefecture. It is decorated with long ribbons woven with silk thread and decorated with silk whiskers. It was used to contain areca in the past, so it has this name and has now become a general decoration.

Hanging box is a kind of large decoration used by Tibetans. It also has the function of protecting the body. It has the shape of Buddha's niche and square diamond pestle and so on.

In some places, bamboo baskets are not only tools for carrying things, but also personal belongings. When going out, people often carry it on the back, even if it is empty. There are large and small baskets. The small ones are typical of the waist baskets of Dai people in Yuanjiang and Xinping. The waist basket is called Yanggan in Dai. It is tied around the waist to hold small items and is also an ornament. The waist basket used by young people are also decorated with wire balls and whiskers, which are extremely exquisite.

Some ethnic groups have beautifully embroidered money bags, which are tied around the waist to put money and articles. Some ethnic groups such as Buyi have reservations. The waist palm made by the Bai nationality is somewhat similar to a money bag. It is also wrapped around the waist, but there is no bag to put things in. It is like a pocket.

Straw hat and bamboo hat are tools for sun protection and rain protection. However, in many places, they are also equipped with delicate embroidered ribbons or silver chains, which can be used as personal accessories when going out. The two kinds of small bamboo hats of Dai people in Xinping and Yuanjiang are mainly decorative. First, it is similar to the blooming termite mushroom. Another is similar to the newly opened termite mushroom, which is decorated on the head, and some are painted.

Dai people in Xinping, Yuanjiang and other places do not wear bamboo hats, but use umbrellas as as decoration, and take them when they go out. Zhuang, Yao, Yi and other nationalities also have similar customs. They also make exquisite umbrella covers, which are usually carried on their backs as ornaments.

Coir raincoat is a rain gear and a special member of the clothing family. There are several kinds of coir raincoats made by Yunnan ethnic people, which are made of mountain grass, palm, bamboo, papyrus and other materials. The materials are different, and the shape of the clothes is also different. The mountain grass is well woven and sewn, which is time-consuming. The broken bamboo silk is connected and stacked layer by layer, so that the rain can not penetrate. The palm is made from the fiber net on its petiole, and the long leaves of the lollipop tree are taken to remove the thorns and dry in the sun, and sewn into a poncho. Each has its own advantages and disadvantages.

The back quilt is used to carry children. It is also a fine handicraft. Because the back quilt is wide, it is suitable for embroidering patterns. People in most places need to embroider flowers in the center and around the back quilt for decoration. As far as Yunnan is concerned, Zhuang people in Wenshan Zhuang and Miao Autonomous Prefecture,

Bai people in Dali Bai Autonomous Prefecture, Yi people in Chuxiong Yi Autonomous Prefecture, Weishan Yi and Hui Autonomous County, Luquan Yi and Miao Autonomous County, Shiping County, Honghe County, Shilin Yi Autonomous County and other places are outstanding in embroidery technology. The back quilts of Miao people are also beautifully embroidered. These back quilts are not only well embroidered, but also have extremely rich cultural connotations.Take a back quilt of Yi nationality in Wuding County as an example. It is embroidered with flowers, pomegranates, birds, butterfly peaches, fish playing with lotus, rabbit pulling radish crab Buddha's hand, mouse grape, magpie climbing plum, chrysanthemum, etc. The upper edge of the back quilt is also embroidered with pine, bamboo and plum. There is a magpie on the plum, and the inscription "plum blossoms are rich and noble, and magpies report peace". The content of Zhuang nationality's back quilt is the same. There are thousands of compositions such as dragon dancing clouds, flowers, birds, fish and butterflies. Each small picture has cultural connotation, and small pictures can form large pictures. The back quilt of Bai nationality is embroidered, highlighting the bright red color. The embroidery is compact and tight, and there are many traditional cultural connotations. Different shapes of back quilts in different places also have high artistic value.

Knives are inseparable from the body of many ethnic groups in Yunnan. They can be used as self-defense, tools, and handicrafts. There are two types of knives: long knives and small knives. The long knives are carried by the Achang, Jingpo, Nu, Dulong and other nationalities. As for Lisu, Nu, and Dulong of Nujiang Lisu Autonomous Prefecture, the iconic long knives are the side knives with only half of the knife shell. The blade is exposed. The knives used by the Jingpo and Achang people are both from the Achang inhabited area. They are similar in shape. The difference lies in the blade. The Achang knife has a blade tip and the Jingpo knife has a flush tip.

Crossbows and arrow bags have also become important personal items in Nujiang area, and some people wear them not only when going up the mountain, but also when going to the market. Every nationality has a shoulder bag, and it is also an important accessory. For Jingpo, Jino, Wa, De'ang, Dai, Xishuangbanna Hani, etc., the accessories and colors of the bags are the same as the clothing, and the local and national characteristics are prominent.

The shoulder bags of the Jingpo people have the same color as the skirt, and the silver bubbles and pendants are similar to clothes, but smaller. The decorative thread of Jino linen satchel is like clothes and trousers. Va shoulder bag has the same tone as the skirt and is decorated with barley. De'ang shoulder bag is like the composition of the clothes, the color of the fabric and the decorations such as pom-poms are close to the clothes. The shoulder bag of Hani people in Xishuangbanna has patterns on one side, embroidered with geometric patterns and other patterns, just like the patterns on the clothes. The bags of Yi nationality are different from clothing, with many varieties and special craftsmanship.

The kudzu fiber bags and mesh bags of Hani, Bulang, and Wenshan Zhuang people in Xishuangbanna and the hemp fiber mesh bags of the Yao, Lisu, and Yi people are also special shoulder bags.

The embroidered bags of Bai people in Dali have a hard shell in color and body, and the embroidery style is unified, forming their own characteristics.

Chapter Six   Clothing Technology

Tie dyeing and batik process

Entering Dali, in the fascinating scenery, the first thing you see is not only the Cangshan Mountain, which is covered with snow all the year round, and the blue and clear Erhai Lake, but also the handmade specialty tie dyed cloth with blue and white flowers.

Tie dyed cloth is a unique craft product of Bai nationality. It can be seen everywhere in Dali's urban and rural streets, shops, stalls, hostels, hotels, hotels and even inside and outside of homes. It is either made into clothes, pants, skirts, vests, coats, sandals, slippers, handkerchiefs, headscarves, bibs, handbags, hanging bags, backpacks, waist bags, or processed into bed sheets, quilt covers, curtains, curtains, wall hangings, computer dust covers, TV covers, or displayed in whole for people to enjoy and buy. Wherever you go, take a look at the clothes and headwear of Bai women, and pay attention to the ornaments worn or worn by other ethnic groups around. You can almost find the shadow of tie dyeing. Tie dyeing not only represents a tradition, but also has become a fashion. When tourists bring tie dye from the local to all parts of the country and even abroad, its steps from ancient times seem to be connected with the contemporary steps. People can not help but marvel at its unique charm.

Tie dyeing is so popular because it is unique and different from other dyed fabrics. It is simple and natural, and the white flowers on the blue background are clear and elegant. It is not publicized at all. It conforms to people's feelings, is close to people's life, and is full of human nature. It is the embodiment of Bai People's good character and interest, such as diligence, simplicity, purity, honesty, kindness, optimism, openness, hospitality and so on. It is said that it was changed by the stream of Cangshan Mountain, or it was woven by fairies to bring it to the world... Wearing it not only looks beautiful, but also represents dexterity and wisdom, and can reflect eternity, kindness and sincerity. The almost ubiquitous tie dyeing has almost become the most dazzling and special cultural symbol and the symbol of national traditional art in Dali.

Tie dyeing is like this, so is the batik of Miao nationality, which is similar to tie dyeing. It is like clusters of brilliant mountain flowers blooming all over the Miao mountain villages of Yunnan Province. As a tourist commodity and cultural specialty, it has already entered the city and increasingly become an important folk art loved by people. Its white and blue mood is like praising Miao Township, praising the natural scenery, showing the character and virtue of Miao people, and emitting the eternal fragrance of soil.

Tie-dyeing.

Tie dyeing is a long-standing fabric dyeing process. Dali people call it a knotted flower cloth and a knotted flower. Knot dyed cloth is also called knot flower cloth and blue flower cloth. Tie dyeing is a name that has gradually become popular in recent years. It is a kind of blue and white dyed cloth that is first "tied" or "sewn" and then diffusely dyed. Because the main production area is Dali, most of the dyers are Bai people. Therefore, people are used to call it Dali tie dyeing. Bai tie dyeing (or knotted flower cloth)is a well-known and popular hand-made tie dyeing product in ancient and modern times.

The tie-dyeing process appeared earlier. In the Song Dynasty Dali National Picture Scroll, there were two warriors who wore cloth crowns on their heads, which were very similar to the traditional tie-dyeing of small groups of white flowers on a blue background. Intuitive records of dyeing used for clothing nearly a thousand years ago, in the temples of the Ming and Qing Dynasties in Dali area, it was found that some Bodhisattva statues had tie-dye fragments, tie-dye scriptures, schoolbags, etc. By the time of the Republic of China, home tie-dye had become very common. , Zhoucheng, Xizhou and other townships, which are famous for the density of tie-dye workshops based on one family, have become famous tie-dye centers in all directions. Many Bai villagers in a hundred Li radius not only go to buy tie dyeing for clothing and accessories, but also go there to learn skills, and then go home to tie and dye themselves. Because of this habit, the tie-dyeing process spreads like wildfire, and the spread is wider and wider, and there are many people who can tie and dye.

Since the founding of the People's Republic of China, with the economic development and the improvement of people's lives, the demand for tie-dyeing has increased year by year, and the output and color varieties of tie-dyeing have been increasing day by day. Especially in the past two decades, under the impetus of reform and opening up, and stimulated by the booming tourism industry that requires a large number of tourism products, tie-dyeing has experienced unprecedented prosperity. Tie-dyeing has jumped out of the single-family production method, and some collective workshops and tie-dye factories with some new technology and equipment have emerged.

In the past, the fabric used for tie-dyeing was entirely made of white cotton soil cloth with a coarse texture that was hand-woven by the Bai people. Nowadays, there are fewer soil cloths, mainly using industrial woven raw white cloth, packaging cloth and other fabrics. The fabric must be highly absorbent, and the processing mainly includes three processes: tying, dyeing, and bleaching. Zhahua, formerly known as tie knot, that is, after the fabric is selected, according to the requirements of the pattern, the fabric is wrinkled, folded, rolled, squeezed and other methods are used to make it into a certain shape, and then stitched with needles and threads. Or tie it, tie it tightly and sew it tightly, so that the fabric becomes a string of "knots". The next step is dip dyeing, which is to put the "knots" fabric into the dye vat and soak it. After a certain period of time, take it out to dry, and then put the fabric into the dye vat for dyeing. Repeat this for several times. The dye is rinsed off, and after drying, the thread is removed, and the "knot" is picked up. The part that is tied and stitched by the thread is not colored, showing a hollow white cloth color, which is the "flower"; the rest is dark blue, That is "ground", so far, a beautiful tie-dye cloth is completed. "Flower" and "ground" are caused by two kinds of dyeing results of color and non-color, and there is often a certain transitional gradient effect between the two, that is, the blue background and the white flower are not formed rigidly. In contrast, the edges of the flowers have faded or thickened color halos caused by stains, making the colors more rich and natural.

In many cases, the effect of this gradual change is artificially controlled by grasping the tightness of the knots. The traditional dyes are mainly plant dyes. In the past, Isatis indigo was mainly used. Because the dyes extracted from Isatis indigo are slow in coloring, low in dyeing efficiency, and the income of the planters is very small. In addition, there are very few growers, so many people use indigo for dyeing. Now, the number of people using chemical dyes is increasing, and the old impregnation method is facing the danger of gradually disappearing. However, the tradition of adhering to the traditional dyeing method, i.e. impregnating with plant dyes, is still maintained in some households, and many tie dyes for personal use are often dyed with plant dyes. This is because people believe that plant dyes such as isatis root or Polygonum leaf have many benefits compared with other chemical dyes according to experience and habits. The most important thing about plant dyes is that they have natural color, fade slowly, do not damage the fabric, and are durable. The cloth dyed with chemical dyes looks dull and bright at first, but after repeated washing and sunlight exposure, it will become lighter and lighter and gradually lose its natural color sense. Vegetable dyes are completely different. With the passage of time, the color of the flower and the bottom will become more and more bright, clear and harmonious. The clothes and skirts sewn with them are vivid and bright in color. They can also bring comfort when wearing and will not cause adverse stimulation to human skin. It is said that dyes such as isatis root also have certain drug effects and are beneficial to people's health. Dali Zhoucheng is the place where Bai tie dyeing is most concentrated, with the largest output and varieties. All the Bai women there can tie flowers, and they use all the time they can to do the sewing and dyeing work for themselves or from other families. In the fields, or in stalls and shops, or even in cars, the baskets are always loaded with cloth, and the flowers are tied at any time. They work with men in dyeing, and the tie dyeing handicrafts produced are colorful and popular, which are deeply loved by the masses.

The tie-dye patterns are dazzling, and the materials mainly include plant patterns (such as flowers, grass, leaves, fruits, etc.), animal patterns (such as insects, birds, animals, fish, etc.), natural patterns (such as sun, moon, stars, clouds, Stone, mountain, water, etc.), character patterns and auspicious patterns. Among them, the most used plant pattern is plant flower and fruit pattern, the first is plum blossom, followed by duckweed flower (broken dot flower), small group chrysanthemum, camellia, frangipani, thorn leaf flower, bean leaf flower, peach flower, ling flower etc. Some patterns have certain meanings, some are purely to express the beauty of natural objects, and some are borrowed from objects to express feelings and reflect the psychological state and ideology of the characters through the depiction of flowers. The most commonly used animal pattern is the butterfly pattern (that is, the butterfly flower). Butterfly is a symbol of motherhood and good weather. Its importance and universal use make it have the status of tie-dye decorative motifs. Large butterflies, small butterflies, single butterflies, double butterflies... All kinds of butterflies can be seen in tie dyeing. Now, the pattern of tie-dye cloth has undergone many new changes. In addition to the traditional patterns, due to the influence of commercialization, some manufacturers have designed and made some products according to the patterns provided by the other party to meet the needs of the orderers, so that non-local traditional patterns at home and abroad continue to appear in tie dyed goods.

Batik.

Batik process and tie dyeing process are like a pair of beautiful and pure sisters in appearance. There are similarities and differences between the two in the production process, which can be said to be the same in different ways and the same in different colors.

The batik of Miao nationality is the most representative in Yunnan. Most of the Miao people in Yunnan live in the mountains. They live by planting, hunting and raising in the mountains. The self-sufficient natural economy occupies an important position. The textile industry is the main sideline in the Miao family and the main source of clothing materials. Miao women have learned to spin and weave with their mothers since they were young. Most of them are skilled in spinning and weaving when they grow up. Therefore, the textile industry has become a social division of labor that is entirely undertaken by women. This situation is still maintained in most Miao areas today. Some ancient textile techniques, such as waist weaving and movable wooden frame weaving, are still widely seen in Miao township. Their textile products are mainly white linen, followed by cotton. The cloth woven by the ancient textile method is the ideal cloth for batik. Its special flavor of rough surface and solid texture is hard to be replaced by other fabrics. This is probably an important reason why the traditional weaving of Miao has been maintained.

The production principle of Miao batik and Bai tie-dye is the same, but the methods are different. Batik generally has the following steps: First, prepare the fabric. The cloth used in Miao batik is self-woven white linen. Miao people cultivate hemp, and every village and household has hemp fields. When the hemp is ripe and harvested, people will tear the hemp skin into thin filaments, wind it into a ball, and then take it to the wood spinning wheel for spinning. After the hemp is spun into a thread, it is bleached with a self-made bleaching agent (usually boiled and soaked with lime or plant ash). The bleached hemp thread is not easy to weave because of its rough texture. Therefore, it is necessary to use a wooden roller to drive the hemp thread on the stone plate with feet to smooth it, After that, put the hemp thread into the water mixed with yellow wax and boil it once to make it soft. Finally, the hemp thread is returned to the hemp ball by the return machine, and the white cloth can be woven by the waist machine. When the cloth is available, the next step is the most important one - to point the wax flower, that is, to draw the pattern to be dyed on the linen with wax. The method of lighting wax flowers is simple. The specific method is to mix beeswax with resin, put it in a small clay bowl, heat it in the fire pool, and after the wax melts, dip the wax with an autocratic wax knife (copper, the knife is about 2 cm wide, shaped like a small axe, connected to a bamboo handle about 15 cm long) and paint the pattern on the cloth. When painting, the thickness of wax should be uniform and moderate. Therefore, the wax should always be placed on the fire pond to keep the temperature and ensure that the dilution of wax meets the requirements of painting. After the wax flower is painted, let it be laid flat or hung up for natural wind curing. After inspection and wax filling for the incomplete part, it can be removed for dyeing. After that, the third step - impregnation. Miao people's dip dyeing raw materials are similar to those of Bai people. They all use plant dyes, and indigo is the most commonly used dye. The production process of Indigo is to cut fresh Polygonum and immerse it in water until it is rotten, then take out its stalks and leaves, add a small amount of lime water and make it ferment. The fermented indigo juice is the dye. When dyeing cloth, put a certain amount of indigo into the water tank to dilute and dissolve, and then put the waxed linen into the water tank for immersion dyeing. Keep stirring in the dyeing tank by hand to make the cloth receive color evenly. After a certain period of time, take out the cloth and dry it, and then put it into the dyeing solution again for dyeing. Repeat this several times. Finally, take out the cloth and put it into boiling water to remove the wax grease, reuse the water to rub and rinse, remove the wax scraps, and dry it. During the dyeing process, the part painted with wax flowers will not be colored into white flowers, and the part not painted with wax flowers will be colored into blue ground or blue flowers,

A beautiful blue and white batik is made.

The batik products of Miao nationality are mainly used to make clothes and skirts, especially pleated skirts for women. Many women's clothing is based on wax dyed cloth, which is added with exquisite and bright embroidery. Secondly, it is used to make bedsheets, quilt covers, headscarves, satchels and other articles. Because of its unique practicality and aesthetic value, batik has gone out of the Miao village and entered the life and social art field of urban people in large numbers. It has been widely used to make various new hats, clothes, skirts, backpacks, satchels, tablecloths and wall decorations such as wall hanging. Among them, batik painting has become a popular new art.

Yunnan Miao batik is fresh and lively, rough and generous, rich in patterns, changeable in patterns, strict in composition, vivid in image and unique in shape, with strong local flavor and national characteristics. Among the batik patterns, geometric patterns are the most common, followed by natural object patterns such as fern grass patterns, bean patterns, sleeping dog patterns, etc. The patterns are unique in shape. Miao people all over the country have different preferences for batik patterns. There are many geometric patterns in the batik of Miao people (flower seedlings and young plants) in southwest Yunnan, and natural objects in the batik of the Miao people in northeast Yunnan occupy more weight. The batik is mainly composed of geometric patterns, tooth patterns, diamond patterns, criss-cross patterns, curved patterns, ribbon patterns, etc. The tone is simple and elegant, the lines are clear, the design is regular, and the overall composition is simple and generous. The continuous and repeated patterns give people a continuous and beautiful feeling, with a strong artistic charm. Batik, which is mainly based on natural patterns, has elegant colors and focuses on expressing a rich, natural and harmonious beauty. The changeable patterns are transformed into lively and lively personalities, which fully reflects Miao people's love for nature and their sincere praise. It is appropriate to compare batik to poetry and painting that exudes the smell of earth, life and the innocence of mountain people.

Artistic features of tie dyeing and batik

When it comes to the artistry of tie dyeing and batik, the first thing to say is its unforgettable blue and the resulting blue mood. Blue is a color full of profound connotation and strong emotional meaning, a color full of rhythm and rhythm, and a color that can call for romantic Association. It is fresh, simple, elegant, stable and quiet. It is like a bridge connecting things and people's feelings. It can lead people's thinking to a broad field, and let people find the feeling of harmony and closeness between people and nature, people and people from various objects full of vitality. At first glance, it seems to be a little rough and monotonous, but it contains rich changes, giving people endless reverie and joy. When people look at, hold or wear tie dyeing or batik, when they see the blue, they will immediately associate it with the blue sky, sea water, autumn mountains, lush forests, etc. It is like poetry written in blue, decorating and beautifying people's lives, allowing people to obtain artistic beauty from its tone.

Tie-dyeing and batik are an ancient handicraft of ethnic minorities in Yunnan. They are the same in color tone and process principle, but there are some differences in the specific expression of artistic style, each with its own characteristics. The tie-dye pattern depends on the condition of the stitches and knots. The key is to work on the hand. Since the stitches can't see the effect of the pattern, the control of the hand makes it a certain chance, and the result of dip dyeing is more casual and natural. Even if the same craftsman makes several pieces of the same pattern, the patterns cannot be exactly the same. The pattern and color of the tie-dye product are directly related to the producer's production experience, skill level, understanding of the pattern forming requirements, the tightness of the tie-dye thread and the number of dyeing times during dyeing. Batik is easier and more artificial than tie-dye in the production and control of patterns. Because batik is the method of first flattening the shape, that is, painting with wax, the effect after dyeing can be predicted intuitively before dyeing. In addition, the part with wax will never be colored, and the production process is like painting, so it is easier to grasp the pattern, so the pattern of wax printing is more artificial. From the effect of the impregnated pattern, the tie dyed pattern is soft, hazy and natural. A large range of non-uniform gradient layers can be produced between the white and blue patterns. The edges of the pattern are fuzzy, as if they were flowers hidden in the water or fog, both bright and implicit. The pattern of traditional batik is generally clearer and more regular than that of tie dyeing. Because wax printing is done by fine hand wax painting, the wax can clearly separate the dyeing liquid from the cloth, so the edge of the dyed pattern is clear, which can produce a sharp contrast with the blue background. Another feature of hand-painted batik flower is that it can give full play to imagination to draw complex and changeable pictures, so that batik presents more delicate and wonderful patterns or rich paintings than tie dyeing. Tie dyeing and batik have their own characteristics. They are the striking varieties of traditional folk crafts in China. Folk and social, local and artistic, they have experienced a long period of time, and they are still prosperous and will always be full of vitality.

Embroidery in national costumes

On festivals and festive days, such scenes can often be seen in ethnic minority areas and Han rural areas: Women's Baotou, hats, clothes, aprons, skirts, pants and shoes are embroidered with various patterns in different numbers and styles; Some men's collars, pockets, satchels, shoes and straw hat belts are also embroidered with colorful patterns; When entertaining at home, embroidered wedding characters are hung in the main room, and bright red embroidered colors are hung under the eaves; The newly married men and women's new houses are covered with lanterns, and the magnificent and dazzling embroidery conveys a joyful atmosphere; The worshippers in the temple offer the Buddha's clothes, red colors, table girdles, flags, etc. embroidered for several months to the incense table of the shrine in the main hall...

Embroidery surrounds people's necessities of life and penetrates into all aspects of people's lives. Because of this technique, people directly draw the various magical beauties observed in the natural world with colorful lines, or through ingenuity, transformation and creation, the ever-changing patterns are preserved and appreciated for a long time. Human images and stories, everything in the world, with this technique, can be presented better than the flat painting viewing effect, enriching life, enabling people to obtain higher material and spiritual enjoyment in life. Embroidery skills are an indispensable part of Yunnan's ethnic costumes and living decorative arts.

Embroidery techniques and materials.

Yunnan ethnic embroidery has a long history. Judging from the patterns on the clothes of the characters on the bronze cultural relics in Shizhai Mountain and other places in Jinning, Yunnan may have had embroidery during the Han and Jin dynasties. In Tang and Song Dynasties, embroidery was very developed. Embroidery has a wide range of uses, and most of them are used to decorate clothes. The situation is roughly the same in ancient and modern times. Some of them are embroidered on the part, and some are embroidered all over the body. The patterns and colors are different for each ethnic group. There are various techniques and styles. There are the following categories:

(1) Cross stitch. Cross stitch is a common embroidery method that uses a needle to pick up the warp and weft threads of the ground (fabric) and thread the thread on the needle under the warp and weft threads, and then embroider various patterns. The cross stitch pattern can be integrated with the ground (fabric). At first glance, it seems to be woven when weaving cloth. Yi people often use this method to embroider patterns such as wrapping the back, burlap, backpack and umbrella cover.

(2) Plain stitch. The straight needle flat embroidery is to use relatively uniform straight needles to embroider, with smooth embroidery surface and no convex and concave feeling. This kind of embroidery method is widely used. It is characterized by the parallel arrangement of needles, including straight, horizontal and oblique rows. The structure is uniform and orderly, without exposure and weight. The straight needle is generally used to embroider flowers, plants, leaves and other pictures. It is often used for large-area backing. A large number of ordinary flower faces are mostly embroidered with this method.

(3) Three dimensional embroidery. The three-dimensional embroidery is also known as the knot wrapped embroidery. One is to create knot like flowers on the flat embroidery screen by the heavy needle method and the stacking method; another is to line the core under the embroidery line (such as wrapped round strips of fine cloth, wrapped cotton thread, brown silk, cow tendon grass, rags, etc.) to make the embroidery surface or embroidery foot texture protrude; there is another kind, which is twisted embroidery or side-by-side embroidery with gold and silver threads. With pearl and jade, the three-dimensional effect is excellent. This embroidery method is commonly used in Bai nationality hanging bags, hanging colors, table girdles, etc.

(4) Stitch embroidery or miscellaneous stitch embroidery. Random needle embroidery refers to the random staggered and overlapping needles in the direction of the needles. It is an irregular and special embroidery method. When embroidering scales, hairs (birds, birds, animals) and flowers, it is used to express their texture and quality. Dynamic is often used.

It is more common in Han, Bai and Yi embroidery.

(5) Appliqué. Appliqué embroidery refers to cutting cloth into a certain pattern, sticking it on the bottom material, and locking and embroidering along the edge on the appliqué. Kucong people of the De'ang, Yi and Lahu nationalities use various pieces of cloth to sew their clothes together, and add embroidery patterns at the seams, which are closer to appliqué embroidery.

(6) Cut-out inner appliqué. Cut-out inner appliqué is similar to appliqué embroidery, in which the embroidery ground (ie, fabric) is cut into a hollow flower shape, and the inner lining is filled with fabric and locked along the edge. The shape resembles an applique, with a concave feel, in contrast to the protruding feeling of appliqué embroidery.

(7) Combination of dyeing and embroidery. Bai nationality uses the tie knot dyed blue cloth (tie dyed cloth) as the base material, and Miao nationality uses the batik dyed cloth as the pleated skirt, and Embroiders it to make the plane and the protruding grain color match each other, resulting in a double-layer picture effect, where elegance and beauty are combined and are harmonious and natural. The peacock dance clothes and elephant feet drum clothes of Dai nationality show the feather feather patterns by means of painting (printing) and embroidery, which are extremely gorgeous. The five Buddha crowns worn by Dongba of Naxi nationality are painted (printed) and embroidered on the base material to draw the image of the gods, which is solid, solemn and beautiful.

(8) Lianwu embroidery. Lianwu embroidery is also called wrapping embroidery, which is a special embroidery method. Lisu People drill holes in seashells, and sew the seashells into embroidered patterns through the edge holes. The stitches of the sewn holes are regular and orderly, mostly in the shape of patterns. Achang people cut the heart of honeysuckle flower stalk into round pieces and carved it into small flower shapes, and sewed it into the embroidered surface like a net along the side of the colorful thread as decoration for clothes and satchels. All over the country, there are luxury pig hair, animal bone grinding products, copper coins, metal rings, colorful stone grinding products, jade snails, clams and so on, which are embroidered on the shoulder bags and crown clothes. The forms are diverse and varied, and the unique three-dimensional sense has a distinctive artistic effect.

The materials used in embroidery are also relatively rich. Generally, cotton cloth, linen brocade, etc. used as embroidery materials are called "ground", while embroidery threads are mostly silk thread, cotton thread, hair, ponytail, assorted thread, etc. Embroidery shows a different pattern than brocade. The pattern of brocade is a flat flower produced by interweaving colorful warp and weft threads, while embroidery is drawn with needles.The thread presents different colors with different needle movement methods, and produces high flowers on the embroidery surface, with a certain concave-convex three-dimensional effect. The tools for embroidery are relatively simple.

Color tone.

Color occupies a special position in embroidery. All ethnic groups in Yunnan have different preferences and advocations for color. In Han and Jin Dynasties, some ethnic groups in Yongchang Commandery (now Baoshan) used orangutan blood to dye Zhu (red felt), covered the dead with white Tonghua cloth, and then sewed clothes for the dead to wear. This shows that these ethnic groups at that time respected Zhu (red) and white. In Tang and Song Dynasties, red, red and purple were the most precious. After the yuan and Ming Dynasties, the literature has made a detailed description of the colors advocated by various nationalities. For example, in Ming and Qing Dynasties, Yi embroidery used blue as the blue sky, red as the bottom, and yellow as the dragon, giving specific meanings to the colors of blue, red, and yellow. This kind of imagination of color caused by vision is a common phenomenon, which originates from the reflection of color on the object image. Through the reflection, refraction, absorption and other functions of light, they interact with each other, change, and cause various associations through people's vision. This kind of conjoined imagination is restricted by material conditions, so there will be love and preference for certain colors, and thus there will be mysterious pursuit. For example, the ancients valued the rare, so they often listed the items that were not easy to embroider and could not be obtained as the top grade. Even black, white, blue, and cyan, which are visible to the eyes, because everyone has them, and everyone loves them, they will unconsciously become customary and become the color of social respect. Up to now, the aesthetic feeling and advocating of color of all ethnic groups have maintained the influence of several traditions handed down from ancient times. For example, Bai people like white and cyan. They often use white and cyan as the base and use the contrasting pure colors (black to white and blue to white) for plain embroidery, because they think that white is a symbol of freshness, purity, brightness, nobility, solemnity and auspiciousness, and cyan represents hope, simplicity, reality and sincere feelings. Green dialogue is clear and aboveboard. Yi, Naxi and Lisu People like black. They think it has the meaning of nobility, diligence, vigor, reliability and sincerity. Therefore, they like black clothes, plain embroidery with white thread on black background and colored embroidery on black background. From the perspective of advocating color and using color habits, the embroidery of ethnic minorities in Yunnan can be divided into three color styles:

(1) Brightness. The color embroidery of Yi, Bai, Naxi, Lisu, Hani and Miao ethnic groups is mostly based on white and black. They like bright red and green. The color contrast is extremely strong and the contrast is large. In addition, neutral colors are mixed with each other to play a transitional or brightening role, making the patterns fresh, eye-catching and vivid. On the color wheel, red and green are complementary colors to each other, and the same amount can be used for embroidery to make red appear more red and green become more green. After the complementary color comparison, the color is bright and saturated. In the strong contrast, it can also make people have a certain illusion in vision. In addition to the contrast between the cold and warm of red and green, the warm color is bright and convex, and the cold color is low and concave, resulting in a three-dimensional undulation feeling of distance and convexity. Moreover, the contrast between the cold and warm, the light and dark levels and the space distance and distance produces a pure and simple aesthetic feeling. This is consistent with the simple, straightforward, vigorous and heroic character of these nationalities.

(2) Simplicity. Bai, Yi, Zhuang and Hani people embroider on the base material of deep or light color with a single color, and make the pattern clear and prominent with the help of color contrast. For example, black-and-white embroidery uses black and white to set off each other. White is based on black, white is highlighted by black reflection, and black is particularly eye-catching by white setting off, thus producing a bright three-dimensional vision. Another example is plain embroidery. The white scarf, handkerchief and apron of the Bai nationality are embroidered on the blue cloth, elegant cloth and Shilin cloth. The purity of the two colors is strong. The hue is bright and dark, and the color is bright and dark.

They are contrasted with each other. They are bright and bright. The Bai people believe that the blue represents the water of Erhai Lake, the white is the snow of Cangshan Mountain, and the aesthetic interest and artistic imagination of the children of the jade Erhai silver Cang seem to be condensed in this kind of embroidery. Plain embroidery is the opposite of colorful embroidery. Although it is only embroidered with a single color, that is, a light and cool thread, it still contains the flavor of "color", that is, the so-called "colorful in plain" artistic conception, which makes people not only feel single flat tone, but will be attracted by the quiet and elegant color. The same is true for Bai and Miao people's blue and white cloth dyed and embroidered.

(3) Light gray. The basic tone of light gray is the same color embroidery or light color thread embroidery. Generally, light colored threads are used to embroider on the same color deep ground or vice versa, and dark tone is more. Hani blue cloth clothes are embroidered with the same color. After washing, the embroidered thread fades and is slightly shallower than the ground. The pattern is indistinct. After washing again and again, the fabric and pattern will fade again and again. After the color gradually weakens, hazy color lines will appear. After the clothes fade, they will be dyed again, and the above situation will be repeated. Yi, Bai, Zhuang, Naxi and other ethnic groups have embroidered with the same dark or light near color patterns on the shallow base. After washing with water, the dark embroidered feet show excellent stains, and the hidden patterns are indistinctly displayed. It seems that the patterns become double layers, and special effects appear. The grayish tone is subtle and mysterious, and the layers are hidden in the mist, which contains the beauty of simplicity.

Pattern types and cultural intercommunication

There are many kinds of embroidery patterns of various nationalities in Yunnan. The themes, shapes and colors of the patterns are so diverse that people are dazzled. On the surface, it seems to be changeable, as if it has magic power. But when we actually analyze it, we will find that most of them are from objective archetypes, life and nature. They are based on various archetypes to create a basic pattern and then gradually evolve, expand and deform. After careful consideration, the prototype basis of each basic pattern can be roughly divided into three categories:

(1) Many of the patterns show the natural objects in the living environment and people's understanding and imagination of them. They permeate people's feelings for nature. These natural objects include:

Plants: ferns, grass vines, kudzu, trees, flowers, crops and fruits. The number of flowers is the most, such as cauliflower, Mushroom Flower, prickly flower, prickly leaf flower, star anise, wild chrysanthemum, jasmine flower, pomegranate flower, chrysanthemum, daffodil, rose, dough flower, lotus, fennel flower, plantain flower and other miscellaneous flowers. There are many embroideries that are purely made of flowers (including buds, flowers, petals and flower stems, branches, leaves, leaf vines and flowerpots, vases, flower baskets and flower windows) or mainly composed of flowers, such as the head scarf, apron, streamer, back, strap, straw hat strap, mouth string bag, shoe upper of the Yi and Bai ethnic groups, Naxi streamer, Lisu dress, Hani sleeve, Dai apron, hat streamer, etc.

Animals: animals can be divided into animals, livestock, fish, insects, birds and birds.

Animals include elephants, tigers, deer, monkeys, rabbits, squirrels, muntjacs, otters, Bobcats (broken faced dogs), jade faced spirits (white nosed Bobcats), leopards, porcupines, etc. and their fur patterns. Miao shawls are embroidered in the color of tiger skin pattern. Yi and Bai People's straps, headbands and streamers are decorated with squirrels or monkeys climbing branches, otters and jade faces. Children wear tiger head embroidery hats, bobcat embroidered hats, etc.

Livestock includes horses, cattle, pigs, sheep, mules, donkeys, dogs, cats, rabbits, etc. or their fur patterns. The upper of the Yi People's shoes is embroidered with sheep in an array. Bai and Yi People's pillow towels, tents and eaves are embroidered with chickens, pigs, cattle, sheep and horses. Thread embroidered cat's head hat, rabbit's hat, dog's head shoes and double dog's belly are popular in some ethnic groups. Dog tooth pattern and cat claw pattern are used by almost all nationalities. Birds and birds mainly include chicken, pheasant, brocade, chicken, Silver Pheasant, peacock, Phoenix, magpie, mountain magpie, sunbird, fire pigeon, emerald, water duck, osprey, white crane, quail, parrot, swallow, heron, mandarin duck, sparrow (Sparrow), turtle dove, etc. or their feather shapes and patterns. Lahu and De'ang ethnic groups embroider chicken claw flowers on clothes, skirts and handbags. Hani people embroidered Silver Pheasant, double chicken in bag and streamer.

The embroidery patterns of Bai and Yi ethnic groups include four Phoenix wearing flowers, two phoenix sunrise, swallow coming to spring, robin, thrush, osprey, parrot, rooster, etc. Colorful embroidered banners such as pine crane, red phoenix, rooster and magpie are hung in the home lobby. A mandarin duck pillow is prepared for marriage, and the husband and wife give each other Yan Shuangfei handkerchief. On the clothes, Lahu people embroidered chicken feet, turtledoves with open mouths, and the De'ang people embroidered chicken feet.

The fish include carp, slender fish, bighead fish, round fish, longan fish, bowfish, mud, loach or fish scale, and fish tail pattern. It is often seen in the colorful embroidery around the waist, streamers, uppers, children's hats and tables. Ethnic groups living near the water area, such as Bai, Dai, Zhuang and Yi, prefer to use such patterns.

Insects include bees, ants, butterflies, dragonflies, cicadas, moths, snails, polypodies, centipedes, water beetles, fireflies, spiders, crickets, water bench worms, beetles, cockroaches, grasshoppers, caterpillars, locusts, or insect back patterns. The embroidery of Jingpo, Dai, Yi and Hani ethnic groups has the deformation of snake and earthworm patterns. There are many nationalities weaving and embroidering bees and butterfly patterns. The butterfly patterns embroidered by the Dai hat band, the Bai belt, the Yi waist band, the Lisu waist band, the Miao embroidery band, and the Achang head bag top decoration are all extremely exquisite. The spider web pattern, centipede pattern and caterpillar foot pattern on the clothes and skirts of Lahu and Va people are simple and natural.

Celestial and natural objects. Celestial bodies and natural patterns include sky, earth, sun, moon, star, cloud, rainbow, water, river, mountain, stone, day, night, etc. the stripes, curved lines and geometric patterns on some ethnic embroidery often have the meaning of heaven, earth and water. The "seven stars of the sun and moon" sheepskin on the back of Naxi Women is embroidered with one sun and one moon, and five stars. The center of the star is sewn with two suede ribbons to symbolize the radiance. The square embroidery worn by Miao people on the back of their necks shows "ancient roads" and "ancient cities" reflected by the sky, the earth, the sun and the moon. Lahu, Va and Hani ethnic groups weave and embroider patterns such as the sun, rainbow, sea water and mountain forests. The sea water pattern of the Lahu ethnic group also shows the reflection of water. Tibetan

Reba (artists) hang five colored embroidery and silk (silk) ribbons on their backs. Each color represents one thing, symbolizing the sky, the earth, the sun, the moon and the sea. Yi and Bai nationalities use colors to set off the patterns of the sun and the moon. The bright is the day and the dark is the night, which means the morning and evening are auspicious. It is said that the children wrapped in the back quilt receive the essence of the sun and the moon and can live a long life.

(2)Embroidery is like writing, painting and singing. It's like offering sacrifices and conveying love. No matter how much it expresses, people, especially native people, are the core of its expression and praise. Therefore, there are many people and their living culture in the patterns, such as:

Human appearance: The human appearance category mostly expresses people's clothing, appearance and various postures of standing, sitting, lying and walking, and also expresses all or part of human limbs and organs.

Activity scenes: the activity scenes mainly show people's clothing, food, housing and transportation, such as Riding Elephants, riding horses, talking, chasing animals, grazing, feeding birds, singing, dancing, racing, and working (picking, harvesting, digging, carrying, planting, and handicrafts). Yi, Naxi and Bai ethnic groups have the pattern of people walking together, the pattern of two people, four people carrying sedans, two people carrying baskets and wearing flowes. Va people embroidered the dancing posture and formation of several people dancing "bang Bei" (song and dance) hand in hand on the side of Baotou with colored thread. Other patterns, such as two people talking with each other, four people looking at plum blossoms, one or several people catching fish, herding sheep, watering flowers and picking fruit, are popular in various ethnic areas.

Text: there are some text patterns in embroidery, such as Wanzi pattern, Shouzi pattern, cross pattern and herringbone pattern, which are interspersed in patterns. In addition, sentences and characters are also common in embroidery around the world, such as the banners in the nave with the characters "Fu", "Lu", "Shou" and "Xi", the curtains, pillows, quilt covers and table covers with the characters "Xi", the words "love is like a pair of swallows on the beam, and the characters are like two mandarin ducks in the pool", "the water shines on the heart, the wind spreads the meaning, and the good flowers are hidden in the thorns." the handkerchiefs that express love as gifts of the opposite sex include four seasons, good luck, Buddha's destiny, long life and prosperity, heaven's official blessings and other scriptures. The minority languages include Tibetan embroidered mantras and scriptures (some are Sanskrit); Yi auspicious language; The prayer scriptures of Xishuangbanna Dai and Dehong Dai. Since the founding of the people's Republic of China, many new contents have emerged in the text embroidery and decoration, such as happiness, unity, fraternity, diligence and thrift, high mountains and long rivers, mutual respect and love, and love for each other. Some directly embroider the words of loving the party and socialism.

Architecture and tools: Bai, Yi and Naxi embroidery includes arch bridges, wooden boats, livestock carts, houses, screen walls, field sheds, Longmen, pavilions, fields, roads and grooves. The "city" embroidered on the back of the Miao people's neck has walls, roads, houses and water wells. Labor tools and daily necessities such as crossbows, arrows, hoes, spinning wheels, sickles, shuttles, darts, forks, gourds, ladles, shuttles, fences, sheds, fishing nets, animal nets, wine containers, Zhai boxes, Ruyi, chessboard, musical instruments, copper guns and other patterns are popular among all ethnic groups and slightly different from each other.

(3)Some patterns do not necessarily have a clear meaning. They are mainly used to set off other patterns or to beautify graphics. These patterns are called decorative patterns, and are mostly composed of pure geometric shapes and free shapes. Each ethnic group has its own unique patterns, such as the straight and curved patterns of Jingpo, Achang, Dai, De'ang and Jino; Pockmarked spots, shuttle needle eyes, presser foot lines, etc. of Bai and Yi Nationalities.

Embroidery patterns are generally constructed in a single pattern, that is, one pattern is an independent pattern. Single patterns are mostly seen on the hat top, apron, streamer head, heel, etc., and there are also composite patterns, i.e. the combination of multiple patterns or the continuous arrangement of single patterns, such as the head scarf, apron, trouser edge and the patterns "Sifenglinchuang" of Bai, Yi and Lisu, etc. In addition, there is another structural pattern, the nested pattern (radial pattern), which is centered on one pattern and extends to the four directions. The common decorations are the Seven Star decoration of the sun and moon of the Naxi nationality, and the Yuanqing Baotou embroidery of women of the Dai and Achang nationalities. Although there are certain rules in the pattern structure, the pattern organization and pattern changes are random from time to time. When embroidering, some first use lime or ink to make a line draft (dark cloth is painted with lime water, and light cloth is traced with ink), while some do not make a sample, but are conceived by heart, and the thread is drawn with the needle, and are embroidered at will in a basic format. Through skillful techniques, we can produce various unexpected and fresh effects and give full play to our wisdom and talents.

Nature and life are the sources of materials for ethnic embroidery patterns in Yunnan. Among various strange and varied patterns, there are profound national historical and cultural traditions, strong mountain life, and the wisdom, sweat and feelings of the people of all ethnic groups. When carefully reading the contents of embroidery works, people can feel the treasure of each nation for its own national culture, which is the main reason why embroidery is colorful, varied and full of vitality. People love embroidery and closely link embroidery with people's activities, which was the case in ancient times and is also the case today. No matter in the dam area or the mountain area, it can be seen at any time that women of all ethnic groups use all their leisure time to pick flowers and embroider flowers at any time. Even when they go to the deep mountains and forests to graze, they do not forget to wear a bandage and sit down and embroider while shouting at the herds. Of course, at this moment, all the beautiful scenery in front of the embroiderer may be left on the fabric with the needle and thread, and the embroidery is used as a watch

Nature and life are the source of materials for Yunnan ethnic embroidery patterns. In all kinds of peculiar and changeable patterns, there are deep national historical and cultural traditions, a strong mountain life flavor, and the wisdom, sweat and feelings of the people of all ethnic groups. When reading the content of embroidery works carefully, people can feel the cherishment of each ethnic group for their own national culture, which is the main reason for the colorful, varied patterns and full of vitality of embroidery. People love embroidery and closely connect embroidery and human activities. It was like this in ancient times and it is also the same today. No matter in the dam area or the mountainous area, women of all ethnic groups can be seen picking flowers and embroidering flowers at any time in their spare time. Even if they go to graze in the deep mountains and dense forests, they don't forget to bring their flower stretchers and sit on the ground while shouting to the herd. embroidered. Of course, at this moment, all the beautiful scenery in front of the embroidery girl may stay on the fabric with the needle and thread, and use embroidery as a carrier to express feelings and beauty, praise the embroidery creator's own beautiful life and the cultural tradition of the nation, praise The nature they depend on for survival. Through patterns and colors, it is subtly generalized and deformed, which embodies simple feelings and beautiful artistic conception, making it always accompany people, making the world of Yunnan always colorful.

Brocade craftsmanship

"Weaving color is called brocade, and weaving plain is called Qi." According to the book of Later Han Dynasty · Biography of the Southwest Yi People, the people in Ailao "know how to dye and embroider... And weave articles like silk brocade." It can be seen that brocade has a long history in Yunnan. Among the ethnic minorities in Yunnan, the famous brocade are Dai brocade and Zhuang brocade. Secondly, the brocade patterns of Jingpo people are rich. Miao, Bulang, Tibetan, Lahu, Achang, Jinuo, Va, Dulong, Lisu, Naxi, Yi and other nationalities also have brocades, but they are not widely used. Although brocade has different weaving skills, patterns and uses among different nationalities, it has unique techniques and beauty.

Brocade is a special art. It not only has special requirements on composition, but also has a variety of techniques. Color matching and decorative layout are all in the artistic conception. The ingeniousness of brocade composition is mainly reflected in the treatment of geometric patterns. A large number of main patterns in geometric patterns are prismatic, square and octagonal. These simple unit patterns, through the displacement of the same shape and the transformation of similar shapes, are very rich, lively and lively without losing their simple beauty. Although some patterns are only some dots and lines, the special effect produced by the pattern design law of "repetition" and "neat beauty", plus the interspersed arrangement of thick and thin lines and the change of width and density, make the patterns more magnificent, colorful and show a rhythmic sense of movement.

Brocade has a wide range of uses, such as making clothes, ornaments, curtain curtains, art decorations or hanging objects, etc. Whatever it is used for, it is pleasing to the eye and can never be tired of seeing. Probably because of the spiritual effect of brocade's characteristics, in ancient Yunnan, brocade was once regarded as the clothing material to distinguish the identity and status of others. In Tang Dynasty (618 - 907), sericulture and spinning were popular in Yunnan, and various silk brocades were produced. Kings and prime ministers made clothes with brocade. All clothes made of vermilion and purple were top-grade and enjoyed by officials. Ordinary people were not allowed to use brocade as clothing. The beauty and value of brocade in people's minds can be seen from this.

The brocade production in Yunnan was once more common and famous in history. In modern times, it was mainly preserved and continued in some minority areas. In the villages and towns of the Dai, Jingpo, De'ang, Buyi, WA, Lahu, Zhuang and Miao Nationalities, there were many brocade experts in the 1950s and 1960s, and the output of brocade was also large. In addition to self-sufficiency, a small number of them had surplus and could be traded in the market. For example, the Dai brocade of Xishuangbanna and Dehong was like this. With the economic development, a large number of industrial textiles have poured into the market, which has impacted and squeezed the traditional manual textile industry. The number of folk brocades has suddenly decreased, and the output has dropped significantly. Since the 1980s, the night weaving noise in some villages with more brocades has become less and less, and the ancient manual brocade technology is facing the danger of gradually disappearing.

Brocade is mainly woven by hand looms. Dai and Zhuang people use wooden frame looms, Jingpo, Va, Bulang, De'ang and Lahu people use waist looms, and Miao people use both looms. Brocade is generally rich in colors and patterns. The patterns are mainly taken from the original shape of the local nature and life, or imitated, or abstracted and refined to form various vivid ornamental patterns with strong local ethnic characteristics.

Dai brocade.

Dai brocade is a handicraft treasure of Dai women. Dai brocade patterns do not blindly imitate nature. Some small insects and flowers that people do not like in life are treated by their skillful hands, or simplified or exaggerated into geometric patterns, which are "integrated" into nature, adding more decorative and more beautiful than the original nature. The traditional Dai brocade patterns include elephants, peacocks, shellfish leaves, trees, figures, and horses. Although they have been abstracted and deformed, they are vivid and concise.

The patterns of Dai brocade are gorgeous and fantastic. The patterns are mostly in the form of animals, flowers, trees, grasses, houses, etc., which are abstracted and transformed into more general geometric patterns, and then staggered to produce a variety of visual effects. In a kind of Dai brocade, the motif of which is dominated by a certain pattern is often used as the title of this brocade. There are several common ways to address Dai brocade according to patterns:

(1) Diamond and octagonal brocade. This is one of the most common brocade widely used in life. It is more common in the Dai villages of Dehong. It consists of continuous or nested rhombus geometric patterns, derived facies groups, or octagonal pattern continuous changing facies groups, intricately superimposed. It is woven with black cotton thread as the warp and colored thread as the weft, and interlaced with flowers and grasses and geometric patterns as the lining. It has many colors, clear layers, bright and rich, solemn and beautiful. There are also diamond and octagonal plain brocades woven in two colors, which are simple and elegant and very durable. In addition, diamond pattern and double bird pattern brocade are also common in Dai brocade, and the frequency of this kind of brocade pattern is high.

(2) Elephant foot pattern brocade. The pattern consists of abstract connected elephant foot patterns. It is made of dark blue, red cotton thread and yellow silk thread. The color is thick and rich, and the texture is thick. It can be used as a quilt for decades.

(3) Peacock brocade.Dai people regard peacocks as a symbol of beauty, auspiciousness and kindness. The Dai brocade woven with peacock patterns has many patterns and fresh and beautiful colors. It is often a wedding wish.

(4) Roof Texture brocade. The roof silhouette pattern brocade woven with the roof of Dai Buddhist temple and its ornaments as the theme is often used to worship the Buddha. The pattern shape is unique and vivid, and the colors are simple and beautiful.

(5) Elephant tower house pattern brocade. Patterns composed of elephant-carrying pagoda rooms, elephant-carrying offering saddle patterns, chanting pavilion patterns and

dedicating style patterns are often used to decorate Buddha flags, and some are also used to decorate sheets, pillows, pillows, hand towels, etc.

(6) Bodhi double bird brocade. The double bird pattern of Bodhi contains auspicious meaning. It represents the bodhi tree and double birds of the Buddha to protect the living beings in the world. It is one of the common patterns of Dai brocade. This kind of brocade is mostly made of cotton and is often used for bed sheets, quilt covers, tablecloths and offerings.

(7) A braided brocade. The winding board is a winding tool, which is a necessary thing for Dai women to carry out textile activities. Some skilled craftsmen carve delicate patterns on the winding board, making the winding board itself a practical and beautiful handicraft. And weaving its beautiful image into Dai brocade reflects the rich imagination and creativity of Dai women.

(8) Brocade with human, horse and boat patterns. The brocade skillfully deforms and combines the images of people, horses and boats into one, which not only shows the real life of people walking on boats and horses, but also contains the artistic conception of the journey to the Buddhist temple. Its composition is distinct and its imagination is unique. It is one of the decorative patterns of the brocade used for Dai Buddhist flags.

(9) Brocade with flower carrying horses and running horses, kneeling horses and leading horses. This kind of brocade pattern is similar to the meaning of the above-mentioned human horse and boat patterns. It shows people's devotion to Buddha by carrying flowers on horses. Since horses are commonly used in life, this kind of brocade is widely used and many people can weave it. Patterns and shapes are often different from person to person when weaving, and there are many changes. The country style and human touch are very strong.

In addition, Dai brocade also has chessboard patterns, flower and bird patterns, elephant patterns, elephant trunk patterns, animal patterns, turtle patterns, crab patterns, etc. as the theme and mixed with other patterns, each with their own personality, elegant and gorgeous.

The unique colors and patterns constitute the unique style of Dai brocade. Its content and form present the picturesque natural beauty of Dai's heaven and earth, mountains and rivers, rivers, forests, flowers and plants, clouds and neon, and the leisurely life scenes of

Dai's poetic clothing, food, housing, transportation, beliefs and customs. Dai brocade is not only an excellent folk art of the Dai people, but also a microcosm of Dai People's life and culture.

Jingpo brocade.

Jingpo brocade is a unique style of Yunnan national brocade. No matter who goes to Jingpo mountain, the most impressive thing is probably the tube skirt sewn with it. It has bright colors and rich patterns. It is dignified and luxurious on the girls, showing an intoxicating artistic effect and is particularly eye-catching.

The purpose of Jingpo brocade is not only to sew barrel-shaped skirts, but also to make handbags, quilt covers, pillows, leg covers, hanging decorations and decorations. Jingpo brocade is made of the same materials as Dai brocade. It is also made of traditional cotton, silk and wool. In weaving technology, it is mainly woven by looms (commonly known as waist looms). The method is that the weaver sits on the ground, one end of the warp yarn is wrapped around the roller and supported by the sole of his foot, or the warp yarn head is tied to the ground pile or room column, and the other end is tied to the waist. The weaver uses a wooden knife to guide the weft and beat the weft. It is commonly known as Juzhi. This is an ancient weaving method. The present weaving method has been improved compared with the previous one. A thread heald device made of fine bamboo or brown is used to lift the warp yarn and form the weaving mouth. The weft insertion and weaving plus the function of a smooth bamboo and wood pick knife make the warp and the warp The ground warp is separated, and the weft threads of various colors are used to pass through it, which can not only weave a 90-degree plain weave, but also weave a diagonal line pattern. Although the effect is slow, the product is strong, natural and gorgeous, which is much better than the ancient weaving method.

The colors of Jingpo brocade are bright like the morning glow, like fireworks, rich, magnificent and extraordinarily luxurious. There are two main types of brocade, one is black ground red brocade, which is the most productive one. The ground is black with rich tones. The pattern is mainly dark red with cinnabar, and other colors are embellished in an appropriate amount to make the contrast between red and black strong, and then matched with bright yellow embellishment. The primary and secondary are clear, the layers are rich, and the tones are strong. Coupled with the special effect of the refraction change caused by the natural texture of cotton and wool and the suede feeling, it is particularly solemn, thick, rich, and unique. The other is that the base is blue or black, and the quality is like cotton. On a large blue or black base, weaving strips of color patterns at appropriate intervals, with black color support, it is elegant and magnificent, with a strong artistic flavor.

The patterns of Jingpo brocade are mostly composed of geometric patterns, wave patterns, continuous patterns, etc., which are more general and abstract.

The barrel skirt of Jingpo people is another style of brocade different from the Dai brocade. It uses a geometric pattern to weave all over the bottom, with red as the keynote and black as the backing, and the pattern contrast is very strong. On the brocade barrel skirts worn by Jingpo women, all the patterns have not only names but also meanings, as the saying goes: "On the tube skirts, the things of the world are woven, and those are the words written by the ancestors." The intelligent Jingpo women, while performing artistic work and carefully calculating the longitude and latitude lines, have mixed rich feelings such as praise for nature, love for life, longing for the future and wishes for love. Therefore, some people say that Jingpo brocade can convey feelings and can be translated into lyric poems praising nature.

Va brocade.

When you see the Va brocade weavers in the Awa mountain, you almost have a feeling: the Va brocade is like a rainbow, beautiful and natural. Each pattern is like a portrayal of the living environment and life interests of the Va compatriots, which is unforgettable. Va brocade has many uses. First, it is used to sew women's tube skirts, in which girls wear colorful and bright colors. The color levels of middle-aged and old women change less, and the tone is also slightly darker. The second is used for making satchels, leggings, covers, bedsheets, etc. Va men and women use Va satchels. When they are hung on the shoulders of men with a single dress tone, they are very rich because of the contrast. In the past, Va women in Cangyuan, Ximeng and Menglian were generally good at weaving Va brocade with various patterns. Later, due to the popularity of industrial textiles, people's demand for brocade has gradually decreased, and the number of people who often weave brocade has gradually decreased.

According to the quality, there are mainly three types of Va brocade. The first type is made of hemp, all of which are woven with hemp thread (called "pole star" in Va language). The pattern is mainly linear, with red and black lines on the white background. The tone and pattern are relatively simple. This is the original type of Va brocade. The second type is cotton brocade (called "Gandai" in Va language). It is mainly woven with cotton yarn. The colors and patterns are changeable. In addition to dyed cotton yarn, the patterns are also interspersed with hemp, wool and imitation gold and silver threads, making the patterns look fresh and brilliant. The third type is mixed brocade, including cotton and linen, cotton and linen silk and cotton wool. Due to the use of a variety of materials, the choice of color and pattern is increased, so the brocade and tone are relatively free.

Wa brocade is woven on a waist loom. The weaver sits on the ground and ties one end of the warp thread to the room pillar (or the tree on the side of the room) and the other end to the wide belt tied to the waist. Pick up or press down the warp yarn with several thin bamboo sticks according to the rules, pick out the weaving hole, guide the weft through the weaving hole with a shuttle, straighten it. Then, the weft thread is tightened with a comb plate passing through the warp thread, and so on. Although the weaving speed is slow, the thread is uniform and the holes are dense, and the quality is quite good.

Elegant and beautiful patterns are a major feature of Va brocade. The most common pattern in the Va brocade pattern is the bird eye flower, which is called "Ai Xing" in Va language. Its basic shape is to add dots in the diamond frame to make the "eyelids" and "eyeballs" complete, and weave them into birds' eyes as if they were open. In addition, there are tiger foot flower (called "Jiang Si Wai" in Va), geometric pattern wrapped flower (called "Guo Ge" in Va), curved frame flower (called "Luo" in Va) and so on.

The Va compatriots have lived in the mountains for generations, and have close relations with forests, springs, birds, wild animals, flowers and fruits. They have established deep feelings with nature, which is also clearly reflected in the brocade. According to the folk saying, the color of brocade is black at the bottom, which means that it is reliable, like relying on the mountains and forests, and is the source of endless food and clothing; The green, red, yellow and purple patterns are also closely related to plants and animals. The pattern of brocade - bird eye flower originates from people's contact and observation with flowers and birds and the resulting favorable impression on them. According to Va old men, the bird is smarter and more capable than the wild animals on the ground, because it has a pair of wings and can go to heaven. The bird's eyes not only represent wisdom, but also a symbol of heaven and earth, the sun and the moon, because it can see far and wide. Therefore it is a symbol of good luck. With it, you can go hunting in the mountains and work in the fields. No matter whether it is sunny or rainy, you can avoid disasters. If human eyes are as bright as freckles, there is nothing to worry about in the world.

The ethnic brocade of Yunnan, like colorful clouds, is gorgeous and colorful, setting off the continuous landscape. The intelligence, wisdom and skills of the people of all ethnic groups have decorated the land of thousands of miles with various colors.

Chapter Seven   Clothing and Culture

To deeply understand clothing culture, we must start with social culture. Clothing is not just a simple shelter from the cold and wind, it also integrates people's memories of history, understanding of society and prospects for the future. It not only reflects people's world outlook and aesthetic taste, but also reflects life standards and social ethics, covering all aspects of social culture.

Age and clothing

Different ages have different social roles and family identities. Age and role variation are basically the same in society, so people of the same age have the same social roles. Age and role differences are often also reflected in clothing, which forms clothing variations corresponding to a certain age group, and thus forms different clothing types. In a person's life, changes in clothing can be divided into infants, children, youth, middle age and old age. The change of clothing color is the most common form of variation. The colors of children's clothing are bright and lively, and the form is simple. Young people's clothing should pay attention to bright colors, smooth lines, and exquisite workmanship, so as to achieve a higher achievement in clothing art. After entering middle age, bright tones are gradually replaced by elegant and light tones, and mature and prudent become the main deportment. The clothing of the elderly is more plain and simple, using dark colors and no longer focusing on beautiful patterns. Clothing style, accessories and matching are the main body of clothing variation in different age groups. Baby clothes are generally simple and rarely reflect the difference between men and women. Light, soft, warm and breathable are the main standards for making clothes, and accessories are rarely used. In addition to embodying the difference between men and women and being different from baby clothes, children's clothing also embodies the love of mothers and pours love for children into clothing. Therefore, children's clothing is finely made, with lively colors and eye-catching.

Another feature of children's clothing is that it contains people's care and expectations for young life, and there are many healthy and beautiful pattern designs such as warding off evil and auspiciousness. Flowers, birds, insects, fish, zodiac and Buddha statues, longevity, happiness, longevity, fortune, lu and other patterns and characters are obvious examples. These patterns are mostly concentrated on shoes, hats and accessories, making children's clothing The cultural connotations and artistic achievements of these parts are particularly outstanding. Most of the childhood period starts from about 3 years old, and it ranges from 13 to 16 years old. The time is longer and the accumulation of clothing is also large. In Yunnan's ethnic costumes, they have a considerable weight in terms of quantity, types and cultural and artistic achievements. In the past, the coming-of-age ceremony was a sacred ceremony held by many ethnic groups, and dressing up was the most important content. Now some ethnic groups still retain the coming-of-age ceremony and do complicated etiquette.People who do not hold ceremonies regard changing clothes as a sign of adulthood. When they reach a certain age, they can change into the clothes of young people. Even if they enter adulthood, they can participate in the social activities of young people. Adult clothing is a mature and fixed clothing style, which can be kept until old.

From children's clothing to adult clothing, in addition to changes in clothing styles, some iconic clothing changes are common in many ethnic groups. Among the Naxi Mosuo people, changing skirts is an important part of a woman's coming-of-age ceremony. Changing the headgear is the most common kind of adult dressing. Jino women should change their hair into a single braid and wrap around their waists, and use adult clothes for their tops and vests; Jino men mainly use caps and changes in the shape of the sun pattern. To reflect that, adult men will replace their childhood hats with Baotou, and the patterns are only rounded, unlike when they were children. As for Dahei Yi branch of Yi people in Luxi, Mile and other places, the adult hat changing is one of many hat-changing ceremonies. Dahei Yi women wear boat-shaped hats when they are young, and then change to crown hats and long-tailed hats, only after the age of 16. They change it to an adult hat and change it after marriage. The baotou on the top plate of Landianyao women is an adult symbol, while the men's cap is replaced by a handkerchief, and the baotou handkerchief is an adult symbol. In Biyue branch of Hani ethnic group in Mojiang, the most important thing for adult women is to make two hats, only one when they are young and two for adults, because adult girls have to participate in social activities, and young men often take their hats away. In order to show love, it is rude not to wear it. In order to prevent the elderly from asking, we can only prepare one in case of emergency. For Yi and Hani people in some parts of Honghe Hani and Yi Autonomous Prefecture, braiding is an important sign of adulthood. Nuomi branch of Hani ethnic group in Yuanjiang will add ornaments called "pijia" in Hani language to their waists when they become adults.

After adulthood, the style of clothing has been fixed and rarely changed, while some ornaments and iconic ornaments are facing new replacement at the two checkpoints of marriage and childbirth. Generally speaking, wedding dress is the highest level of youth clothing and the peak of clothing art. It still belongs to youth clothing. After marriage or after childbirth, the role and status have changed from the status of a young single aristocrat to another role. After marriage, women have to make certain marks on their clothes to distinguish them from unmarried young people, so as to suit their identity and avoid unnecessary troubles. Most of these changes are headgear changes, some of which are completely replaced, such as hats being replaced by headbands, Baotou, etc., and some only partial changes, mainly changing the shape of the headgear, or removing some parts. Dai people in Maguan County, Dai people in Shiping County, and Hongtouyao, a branch of Yao people in Jinping Miao, Yao and Dai Autonomous County, are examples of such changes. Married or unmarried, the headdress dressing method is obviously different. You can tell at a glance. Awu women of the Yi nationality changed the cockscomb hat before marriage to the Lele hat after marriage, and the change in the headgear of the Sani women also belongs to this scope. Dehong Dai changed from trousers to skirts, which is also a way of identification.

The change of hair style is also a method. There are many kinds of changes, such as changing the hair into a bun, changing the braid, changing the single braid into a double braid, changing the braid into a bun, or changing the style of the hair bun. Such changes are common in most ethnic groups.

Belts and other accessories can also be used to identify married women. Women of Baihong branch of Hani ethnic group have a specially made belt to mark their identity after marriage. The Hani ethnic group called it "Pici". Yi Che's newly wed women use a special bamboo hat to show her identity to passers-by.

The collection and reduction of jewelry is also an important change in the clothing of married women. The jewelry worn when young is gradually reduced after marriage, and is collected for future generations. Married women have too many accessories, which are inevitably ostentatious and inconsistent with their own identity.

The clothing of the elderly is simpler than that of the middle-aged. It pays attention to practicality and does not emphasize fancy. The criteria for entering the old age are not all based on age, but also on personal status and identity changes. Once a child has been born, the children and grandchildren should be around the knee. Even if they are not too old, their clothes and behavior should be coordinated with their identity. In some places, it is a common example that the special signs of the clothing of the elderly can wear a silk shroud. Yi people in Niujie Township, Shiping County have women with grandchildren. In the past, they also wore a special headscarf to distinguish them from middle-aged people.

Wedding and funeral etiquette and clothing

Wedding dress and funeral dress have special significance in dress culture. Wedding is the most glorious and happy moment in one's life. Everyone should appear at their wedding with the best appearance. A clever girl usually takes a long time to prepare her wedding dress, so that she can show her beauty and show her exquisite craftsmanship at the wedding. The wife's family also attaches great importance to grooming the bride, and will do its best to decorate the bride, not only to show the love of the girl, but also to show the strength of the family. Therefore, in the past, the bride's clothes and groom's clothes were extremely luxurious and exaggerated. In particular, the bride's clothes were simply exquisite handicrafts, representing the highest level of everyone, and the matching of accessories was the most complete.

The bride's handkerchief is a dress that a woman uses only once in her life. It was originally from Han nationality. Many ethnic groups have bridal handkerchiefs, some of which are only red cloth, while others can embroider some flowers. They are no longer popular and have gradually withdrawn from the historical stage.

Funeral rites require special costumes. One kind is to let the deceased wear a shroud for the coffin, and the other is for children and grandchildren who wear linen and filial piety. Generally, the shroud should be prepared in advance, and most of them are sewn with silk cloth. In some places, satin is forbidden, because of the satin sound which means that there are no descendants. Some of the patterns of the shroud are the same as those of ordinary old people's clothing, and some have special styles. Generally, embroidery is not emphasized. The shroud is the same as ordinary clothes. Clothes, pants, shoes and hats should be available. There are different customs in different places and the layers of clothes are different. Some places require nine for men and seven for women. Some places count several pieces, and are proud of many. Some places regard wedding clothes and birthday clothes as a whole and make them specially. They only wear them twice in their lives, once at the wedding and once after death.

Filial piety clothes are only worn during the period of filial piety, and are discarded after the ceremony. The production is relatively simple. There are clothes, trousers, shoes, handkerchiefs, etc., which are usually made of white cloth. Only those who avoid white filial piety can use other colors. Filial attire is limited to the deceased's sons and daughters-in-law and other filial sons, while the rest of the funeral attendees only wear filial piety and no decoration.

Love token

Some specific items in clothing, accessories and accessories were often used as love tokens to convey emotions in the past. Some are gifts given to each other when their feelings reach a certain level, and some are tokens of affection. In Yi nationality in the southern border area of Shiping County and Jianshui County, the young women will make clothes and pants to send off lovers, while Hani Lami people in Luchun County have special clothes for lovers. In Kado branch of Hani ethnic group, young women in some places embroider a special flower on the lower right corner of the apron, called Yuehua, which is a sign of single girls, waiting for a lover to take this flower. Picking flowers to show love and starting to fall in love will make people laugh if no one cares about it for a long time. The embroidered flowers include camellia, peony, Rhododendron and so on. Yi people in Niujie township of Shiping County used to send flower shoes embroidered by women to their lovers as keepsakes. Hani people in Yuanjiang Hani, Yi and Dai Autonomous County and other places have sent girdles to pledge their love. Yi people also use the girdles as love tokens, and the girdles are beautifully embroidered. In some places, Kucong people of Lahu Nationality use yellow, green, blue, white and red as keepsakes for men. Men usually send back silk thread, bracelets or other handicrafts. Men of the De'ang ethnic group sent a waist band to the woman as a token. Jino men gave their own lunch boxes, betel nut boxes, needle and thread boxes to the women, and the women gave back belts, straw hat belts, leggings, knife ropes and other things as gifts.

The embroidered coat is a love token of the Yi and Hani nationalities in some areas. Yi people have two kinds of small bags for men, one for gunpowder and the other for money. After sending it, the woman will test the man to see if he can use the money in the bag to make profits and use gunpowder and other things to obtain prey. This custom is mainly popular in Dehong Dai and Jingpo Autonomous Prefecture.

Festivals and costumes

The festival is a good opportunity for the people of all ethnic groups to show their skills. People should put on their most beautiful clothes and hang all kinds of beautiful ornaments. The festival often becomes a dress match after another. The torch festival of Yi people, the march street of Bai people, Munaozongge of Jingpo people, the flower mountain festival of Miao people, the water sprinkling festival of Dai people, the March 3rd of Zhuang people, and the knife pole festival of Lisu People are all clothing festivals that compete for beauty. There are also some special requirements for clothing during the festival. Dai people in Xishuangbanna only tie red, yellow and white cloth strips during the water sprinkling Festival, and do not wrap head handkerchiefs. Yi people in Funing County celebrate the palace dancing Festival. The palace head clothes and the palace master clothes are specially designed and used, and they do not wear them at ordinary times. Yi people in Mile city only use grass leaves and wood leaves to cover their bodies during the fire ceremony. Jingpo people do not wear cloth clothes during the soul burial ceremony. In addition to the color painting, they also add leaves and other things to cover their bodies. This kind of activity has both the meaning of tracing ancient rituals and the meaning of exorcising evil and evil, and it is only limited to men to participate, which is harmless.

Many ethnic groups divide clothing into two types: casual clothes and dress clothes. The casual clothes are usually worn during labor and leisure, and the dress clothes are specially prepared for important occasions such as festivals. In some places, the difference between casual clothes and dress clothes is only in the presence or absence of ornaments. Therefore, to see the traditional and authentic national costumes, festivals are a rare opportunity. Not only are they neatly dressed, but there are many people, and all kinds of clothes will be displayed during the festival.

When it comes to clothing display, Yi people in the Tanhua Mountain area of Dayao County really have a veritable clothing display festival. Girls from all villages and villages can show off their intelligence and beauty during the festival, which can not only win the applause of the crowd, but also win the love and pursuit of young men. The festival is now well known as the Festival of Suits. Costumes and dance costumes are a special kind of costume that can often be seen in festivals. Tibetan, Dai, Zhuang, Yi, Bai and many other ethnic groups have opera troupes with unique costumes in groups. The dragon and lion dances of Yi and Bai nationalities, the peacock dance of Dai nationality, and the Guansuo opera of Han nationality are also held during the festival, with unique costumes and masks. These occupy a certain position in Yunnan's ethnic costumes.

Costumes, legends and stories

In many ethnic groups in Yunnan, clothing is a cultural carrier. A simple embroidered piece or a simple pattern may have a moving story and an unusual origin. There are two sources of stories and legends contained in costumes. One is to imitate the meaning according to the shape, and attach a story as an explanation according to the image of the pattern. The other is to record events with pictures, showing historical legends and events with certain patterns. Divided by content, there are several kinds of love Memorial stories, auspicious legends, rescue Memorial, gender relations, memorial of ancestors' deeds, divine gifts, ethics stories, imitation and pictographs, historical events and taboo stories.

The origin of Kaku Baotou of Jingpo nationality, the origin of Kucong people's ribbons, the origin of Sani people's headgear, the origin of Zhuang people's waistband, and the origin of Jino people's sunflower, etc. all attribute the origin of a part of clothing to a beautiful love story, and regard it as a symbol of auspicious and beautiful.

The legend of the three tailed snail ornaments of Bulang people, the peacock hat of the

Zhuang people, and the cockscomb hat of Yi and Hani people belong to the mascot type. It is said that there was a magical three tailed snail, which turned a girl of Bulang nationality into a stunning beauty. The skin color on her face can change three times with different light and weather in a day. Therefore, it is called "Nansanpiao", which means a beauty that changes three times in a day. Because of such a legend, later generations regard the three tailed snail as a divine thing that can make people beautiful and make ornaments to decorate themselves. The peacock hat of Zhuang children is because the peacock pecked at the snake and saved a child named Di Mei. The parents thought that the peacock was an auspicious thing and could avoid disasters, so they embroidered it into a hat for her to wear. Later generations imitated it and it became a fashion. Many branches of the Yi people wear chicken crowns, which give it auspicious and beautiful meanings. It is said that the A-wu branch living in Mile, Luxi and other places wore chicken crowns because the chickens had killed the centipedes that killed people and saved the ancestors. They thought it was a good thing and decorated it on their heads. It is said that the chicken and the centipede make friends. The centipede borrows the chicken's horn to be a guest. Once the chicken borrows it, it does not return it, and forms a hatred. When the chicken sees the centipede, it chases and pecks, becoming its natural enemy. After Awu village was harassed by the centipede essence, it was unbearable. It was occasionally told that if the chicken could conquer the centipede, it would raise a large number of chickens. The chicken pecked the centipede essence, and people had to live in peace. So it was used as decoration. Hani people in Leyu township of Honghe County say that the chicken is the enemy of the devil. Therefore, the image of the chicken is made into a chicken hat, which is worn on the head and regarded as a mascot.

Some ornaments are attributed to the souvenirs of the rescuers, because they have rescued the ancestors, they are regarded as mascots, and passed down from generation to generation. The boat-shaped shoes of the Dai people have been traced back to the idea of ferrying people through water, and were transformed into shoes. The Wa big hand cuff is said to be inspired by earphones avoiding human and bears. The Hani people in Honghe and other places wear clogs, and it is also said that the ancestors used clogs to avoid the enemy, and later generations follow the custom. Some people put the Hani people's clothes on black, which is also said that the ancestors once avoided the pursuit of the enemy because of the black clothes, and the later generations followed the custom. Some Yi people in Weishan, Nanjian, Xiangyun, Midu and other places have a circular cloth or felt ornament on the back waist, with two decorative dots similar to eyes, which are said to represent spiders. Legend has it that the ancestors were chased by the enemy and hid in the cave. As soon as they entered the cave, a spider built a web at the entrance of the cave. The enemy was unable to enter the cave and searched and was saved. In order to thank the spider for saving his life, he embroidered it on the cloth as an ornament. It has been handed down to the present.

The rivalry between the sexes has a long history, and this kind of content is also infiltrated into clothing. In the legends of Bai people, the reason for having an apron is that women are too shrewd and the division of labor and balance between the sexes will be broken. According to the legend of Pumi people, women were smart but men were stupid. Immortals attributed the stupidity of men to women being too smart, so they made women wear jewelry, distracted by playthings, and blocked their minds, so men became smart. In contrast, the legends of Hani people in Yuanyang County praised the intelligence of women, and described the pins on each person's chest as trophies of wits with men. Men appreciated women's intelligence and gave away pins as souvenirs.

Some clothing parts are said to be the remains of an accident. According to the legend of the Weishan area, Baotou began with the battle wound of piroge. Kucong people said that when the flood was flooding, people took shelter in the gourd. After the flood, the God of heaven used a knife to split the gourd and let people out. As a result, the head of Kucong's ancestors was injured, so they had to wrap it with cloth, which became the beginning of Baotou. Dai nationality wears white Baotou, which is said to be derived from the filial piety of the prince, whose ancestors sacrificed to protect women and children. Liangshan Yi men have horn shaped decorations on their heads, which are called Heaven Bodhisattva. It is said that it came from a man named Ali Biri. Because he ate dragon meat, he grew dragon horns and scales. He could only be wrapped with cloth. Later generations made the same dress in memory of him. According to the legend of Nu nationality, in ancient times, there was an emperor who had long horns on his head and covered it with cloth. Because he was an emperor, later generations believed that Baotou might be the source of power, thinking that it was auspicious.

The origin of some clothing components is attributed to God's gift. The Phoenix hat of the Bai nationality is said to be a gift from the Phoenix to mortals. The seven stars of the sun and the moon on the Seven Star shawl of the Naxi women are the sun, the moon and the stars picked by more than three gods to praise the heroine who fought against the drought devil. They were inlaid on her top Yang shirt. Only then did they have the clothing of Naxi people wearing the stars and wearing the moon. It is said that later generations of Naxi women inherited the old system of the Yinggu Yang shirt.

Clothes with short front and long back are popular clothing of many ethnic groups, but in Pumi legends, they have become a testimony to an ethical koan. Legend has it that there are two brothers and sisters who have not seen each other for many years. One day, the younger brother's prey escaped and chased to a place. The family met him and carried him back to feed him. The younger brother chased him and had a dispute with the male owner. The younger brother is ashamed of the elder sister's family. Although his elder sister repeatedly discouraged him from keeping him, he still cut his sleeves and left, leaving behind this strange costume with short front and long back, which shows the ethical concept of a strong and upright nation, and is of great educational significance.

Some clothing patterns are a kind of imitation, or are classified by shape, and are said to be from a similar item. The beautiful Jingpo barrel skirt is said to be woven from the pattern of the beautiful Baban bird. The ribbons on the headgear of the Sani are said to be learning seven-color rainbows. According to legends of Hei Yi costumes in Mile, Luxi and other places, the hat is made from the pot, the skirt is made from the umbrella, and the clothes are made from the robe, so the hat is like a pot and the skirt is like a boneless umbrella.

In the history of graphics, Miao costumes are the most distinctive. The shawls or the back of the clothes of Miao people are embroidered with many different patterns such as squares, arrows and forks. It is explained that they are the patterns depicting the ancestral land of Miao people, representing rivers, cities, fields, official seals and some events involving the migration of Miao people. Miao costumes in Zhaotong, Kunming, Chuxiong and other places still generally retain such patterns. Daban Yao in Funing County has 12 plum blossoms embroidered on the waist, which is said to represent the 12 surnames of Yao people. In the headdress of the indigo Yao in Xichou, Malipo and other places, there is a ribbon called Luobai. The embroidered pattern is said to depict the migration history of Yao people.

Barefoot was a common phenomenon in Yunnan history, and shoes were gradually popular in modern times. Among Hani people in the Nanuo area of Yuanjiang Hani, Yi and Dai Autonomous County, it is said that the reason for not wearing shoes is the ancient rule formed by the taboo of wearing shoes. According to legend, the daughter of the Han official married Hani King and hid the map information in her shoes and gave it to the Han official. Hani people were defeated and subjected to slavery. Hani people swore never to wear shoes, and wearing shoes became a taboo.

Apparel and Text

Words are also important ornaments of clothing. Auspicious and beautiful words are often embroidered on clothes and hats or carved on ornaments. In the Han nationality and some ethnic minorities, the words "Lu", "Fu", "Xi", "Shou" and so on are common on clothes and ornaments, while the words "longevity and life-saving" are carved on the chest plate specially for children, playing the role of Amulet.

The clothing of the Tuliao branch of the Zhuang nationality is also embroidered with text, mostly in sentences, on the headband and the chest and back of the clothes. There are Chinese sentences, such as unite, send songs and dances with passion, flowers for spring, improve thinking and so on. There are also Chinese characters that record Zhuang sounds.

Yao people usually embroider characters on Luobai, umbrella cover and children's hat. For example, on one umbrella cover, there are words such as "gold", "water" and "boarding". Tibetans engrave Tibetan language on bracelets and other ornaments.

Auspicious patterns in clothing

There are many patterns in Yunnan ethnic costumes, including auspicious patterns, animal and plant patterns, natural object patterns and artificial image patterns, as well as geometric patterns.

The auspicious patterns in Yunnan ethnic costumes include animal patterns, figure patterns, plant patterns and other complicated patterns. The origin of auspicious patterns is folk belief. The auspicious patterns popular among various ethnic groups in Yunnan include local traditional patterns and patterns gradually formed by beliefs from other places. Patterns from other places are mostly from popular patterns since Ming and Qing Dynasties. The sources of mascots are complex, some from historical legends, some from metaphors, some from pleasant images, and some from homonymy.

The auspicious examples generally accepted by the Chinese people include peace and prosperity, more children and more happiness, harmony in marriage, long life together, honesty and promotion of officials. This kind of theme is common in Yunnan ethnic costumes. Besides Han nationality, many ethnic groups such as Yi, Bai and Zhuang also have this kind of pattern.

Pomegranate, melon, gourd, mouse, grape and other patterns symbolize many children, peach and Ganoderma show longevity, bergamot and bat are borrowed as the word of fortune. Peony symbolizes wealth, money shape shows wealth, lotus shows honesty, lotus root represents even, and Lily shows a hundred years of love, etc. All can be seen embroidered on the waist, back quilt, clothes and hats.

Pines and cypresses are evergreen, and they do not change color in winter. They are regarded as sincere gentlemen. The bearing and style of bamboo, orchid and plum are also praised by people. Chrysanthemum is regarded as a reclusive gentleman of flowers. Pines, bamboos and plums are called the three friends of the year, while plum, orchid, bamboo and chrysanthemum are called the four gentlemen. This is a traditional viewpoint, which also affects the ethnic costumes of Yunnan and it is a common theme.

Dragons, phoenixes and unicorns are common in China and in clothing in Yunnan. Group pictures such as butterfly love flowers, magpie climbing branches, fish playing with lotus and so on are also common patterns.

The patterns and shapes of the twelve zodiac animals are commonly seen in jewelry, shoes and hats, especially children's shoes. They are mainly popular in Han, Yi, Bai and other ethnic groups. The figure patterns are mostly Bai costumes. The satchel and back quilt are often decorated with figures.

Yi and Bai nationalities in Dali will put grain seeds into sachets as mascots for children to wear.

Famous flowers in Yunnan, such as camellia and Lantana, are also regarded as mascots and embroidered on clothing. In addition to dragons, phoenixes and unicorns, the common auspicious animals are lions, tigers, peacocks, silver pheasants, golden pheasants, parrots, cranes, herons, eagles, geese, chickens, ducks, geese, magpies, elephants, carp, frogs, mandarin ducks, crabs and butterflies.

Many patterns in Yunnan ethnic costumes use animals and plants as the base to consider their visual effects. There are many plants selected, some of which are deliberately imitated, and some of which are identified by similarity. If the pattern looks like something, it is regarded as the object. The common ones are tree shape, chicken fir, yellow bubble, duckweed, fern leaf, fern bud, bamboo shoot platform, sesame shell, pepper, willow leaf, etc. Common animal patterns include cicadas, ants, wild ducks, chicken crowns, red deer, dragonflies, snakes, dove tails, beehives, turtles, centipedes, etc. There are clothing patterns and ornaments.

There are many patterns of natural objects used for clothing by various ethnic groups in Yunnan, such as sun, moon, stars, snowflakes, clouds, waves, flames, water, mountains and stones. The artificial products seen in clothing are usually made into small models for ornaments or images. The common ones are lanterns, tops, arrows, knives, spears, swords, bells, bells, bottles, flower baskets, syringes, ear digging, shuttles, pots, houses, and so on.

The common geometric patterns in Yunnan ethnic costumes include square, triangle, trapezoid, rhombus, cross, circle, coil, well, T-shaped pattern, ripple, X-shaped pattern, triangle puzzle, rhombus puzzle, line mosaic, figure 8, mountain shape and other patterns. There are many patterns, and the matching is complex. The patterns are colorful and extremely attractive.

Chapter Eight   Protection and Inheritance of Yunnan Ethnic Costumes

Clothing is a kind of "historical symbol". The inheritance of clothing from generation to generation also means the transmission of culture from generation to generation. National costume itself is a kind of national culture with a long history. It condenses the national aesthetic consciousness, decorative art and ethnic characteristics, and gathers the regional, historical and national flavor. When clothes and ornaments are combined, they give complex national group consciousness and social and cultural functions on the basis of their simple functions of cold protection and warmth preservation, and become a component of social relations, thus showing the artistic beauty and the creativity of national spirit. From the perspective of economics, national costume culture is a valuable resource, which has important value for the development of regional tourism industry, cultural industry and regional social economy.

With the increasing popularity of modern social civilization, the role of reflecting and preserving history through costumes is no longer so important. Some minority costumes have also begun to show a trend of simplicity and practicality, and the aesthetic intention is more obvious, which will inevitably lead to the loss of the national character of costumes. In the 1980s, when we visited the ethnic minority areas in Yunnan, we saw men, women, old and young wearing ethnic costumes working in the fields to go to the market, participating in ethnic festivals, singing and dancing, reflecting the strong ethnic flavor and local characteristics. After the 1990s, it was also seen that middle-aged and old women and children wore national costumes. After 2000, only old women wore national costumes. Today, when we advocate protecting the environment, returning to nature and protecting the traditional cultural heritage of ethnic minorities, the living habits and costumes of ethnic minorities have gradually changed. If this problem can not attract enough attention from the society, in a few generations, the minority dress culture will no longer exist, which will be a great regret in the history of human culture.

Due to the complex and diverse geographical and three-dimensional climate factors in Yunnan Province, people migrate and settle down, and the cultures of various ethnic groups interact with each other for a long time, gradually forming the unique lifestyle and cultural connotation of the ethnic groups. National costumes have also formed a set of traditional styles with profound cultural deposits of the nation. In the past, most ethnic minorities lived in remote areas and lived a self-sufficient small-scale peasant economic life. They were rarely affected by foreign culture, and their pace of life was relatively slow. In some areas, the economy was relatively backward. After the costumes form their own national characteristics, the general changes are relatively small. After the reform and opening up, great changes have taken place in social life. With convenient transportation and the arrival of the digital information age, the space distance of the earth has been shortened, and there has been a big collision of different cultures. These conditions have affected the life style and aesthetic psychology of Yunnan ethnic minorities.

The clothes worn by urban people are diversified and personalized, simple and easy to wear, and reflect the aesthetic consciousness of the times. Most clothing is the product of industrial production, so it is economical. People do not make clothing, but only buy clothing. Most of the traditional costumes of ethnic minorities are made by themselves, and some are made by local tailors. The production process has been handed down from generation to generation. Miao costumes are covered with patterns and embroidery. It takes one year, at least half a year, to make a suit. Some ethnic costumes are also decorated with inorganic materials, such as jade, jadeite, gemstone, agate, turquoise, gold, silver, tin and copper. There are also artificial products, such as ceramics, enamel, glass and so on. The cost of making a suit of clothing is also very high. The collection and processing of raw materials, spinning, weaving, cutting, sewing and embroidery are all carried out after work. Few women can do needlework at home instead of doing farm work and housework. These are laborious and time-consuming labor. After a suit of clothing is made, it is also complicated to wear. Such as packing the head, tying the belt, buttoning, wearing leggings, wearing accessories, and knitting hair.

At present, the middle-aged and young people of ethnic minorities receive the education of modern culture and have more contact with the outside world. They are influenced by the lifestyle of urban people. All these have a subtle impact on their life and aesthetic concepts. The ethnic minorities living and working with the Han nationality have gradually changed their own national customs and neglected their own unique national customs and culture. If we do not adopt the active protection mechanism, it will be quite difficult to restore the national costume culture in the end.

This requires the cultural management departments of the state and local governments to establish and improve an effective active protection mechanism, and also requires the active cooperation and participation of the people.

(1) Establish self-esteem and self-confidence to protect the national clothing culture. The formation of the national costume culture has its own historical origin and rationality. It is also an important part of modern multi-ethnic culture and a symbol of the vitality of the nation. Then, from the perspective of folklore, only by wearing national costumes can we establish a sense of identity and self-esteem for our nation, and know the value of self-development, perfection and independent existence by absorbing the advantages of foreign cultures while protecting our national cultural traditions.

(2) The collection of national costumes by individuals and museums is a way to protect national costume culture. Ethnic minority autonomous provinces and counties have established museums of their own culture. The collection of national cultural relics in local museums should be carefully sorted out and classified. The collection of national costumes should have distinct national characteristics and have the effect of original flavor. The clothing exhibition and display design should achieve a coordinated and unified style. The cultural relics value, practical value and artistic value of national costumes should be studied and introduced to the audience. However, the display effect of national costumes in some local museums is not very ideal. To realize the active protection, it needs the guidance and support of the government, so that the ethnic minorities can continue to wear national costumes in their daily life, and have a love for their own national costumes, and realize that this is also a unique part of modern culture.

(3) The folk and government regularly organize exhibitions of ethnic minority costume

culture not only in the local area, but also at home and even abroad, so that units and people of insight can make acquisitions. Some collectors in the world are very interested in ethnic costumes with originality and strong artistic effect, and their bids are not low. Some people in China also like to collect national costumes as decorations or collectibles on the main indoor wall. Organize experts to demonstrate the national costumes with collection value and market prospect, determine their artistic value and economic value, make them have a broad domestic and international market and professional national costumes manufacturers in order tomake the production process of national costumes continue.

(4) A competition to display self-made national costumes was held and awards were awarded to encourage and carry forward the spirit of traditional costume art. Folk and government organizations celebrate the traditional festivals of ethnic minorities, and they are required to wear their own costumes to show their characteristics.

(5) Found ethnic minority clothing art magazine to study its historical origin, artistic effect and cultural value. Introduce a variety of clothing cultural activities and conduct academic exchanges of national clothing culture. Provide policy guidance for the protection of active national costume art. Guide the commercial circulation of national costumes. Discuss the sustainable development of national costumes. Enhance people's consciousness of protecting the clothing culture of ethnic minorities and enrich people's clothing cultural life.

(6) Developing regional tourism economy is an important means to protect and develop the clothing culture of ethnic minorities. With the development of modern economy, urban residents like to use holidays to go to rural "farmhouse" to cultivate their body and mind. Some people also like to collect ethnic minority costumes, embroidery, ornaments, handbags and handicrafts as souvenirs.

The scenic spots, unique way of life, rough ethnic songs and dances, distinctive architecture and costume culture, and local specialties in ethnic minority areas are all important resources for the development of tourism economy. These all require the local ethnic minorities to dig, organize, develop, and add high-quality services to attract tourists. Wearing ethnic costumes at ordinary times can increase the local style, and the embroidery patterns and pendants on the body can also be arranged and mounted, and sold as tourist products.

Yunnan attracts the world with its magical mountain topography, diverse three-dimensional climate, magical human landscape, and colorful minority images. These are the innate advantages of our province to develop tourism, making Yunnan a major tourism province. However, under the impact of the modern fast-paced life and the wave of economic development, the ethnic minority costume culture in our province has been lost. The government and people should immediately start to establish an active mechanism to protect the ethnic costume culture, and formulate a strategy for the sustainable development of the ethnic costume culture. The costumes of ethnic minorities themselves need to be developed and perfected, so that they can be enriched into the long river of the culture of the times, so that they can be colorful, lively, and coexist with multiculturalism, so as to show the tenacious vitality of the ethnic group.